Joseph Collins

The genesis and dissolution of the faculty of speech

a clinical and psychological study of aphasia

Joseph Collins

The genesis and dissolution of the faculty of speech
a clinical and psychological study of aphasia

ISBN/EAN: 9783744751339

Printed in Europe, USA, Canada, Australia, Japan

Cover: Foto ©berggeist007 / pixelio.de

More available books at **www.hansebooks.com**

THE PROFESSOR MAKES A PROMISE TO LAWRENCE.

YUSSUF THE GUIDE:

BEING

THE STRANGE STORY OF THE TRAVELS IN ASIA MINOR OF
BURNE THE LAWYER, PRESTON THE PROFESSOR,
AND LAWRENCE THE SICK.

BY

GEO. MANVILLE FENN,

Author of "In the King's Name;" "The Golden Magnet;" "Brownsmith's Boy;" "Nat the
Naturalist;" "Bunyip Land;" "Menhardoc;" "Patience Wins;" &c.

WITH EIGHT FULL-PAGE ILLUSTRATIONS
BY JOHN SCHÖNBERG.

Lucem Libris Disseminamus

LONDON:
BLACKIE & SON, 49 & 50 OLD BAILEY, E.C.
GLASGOW, EDINBURGH, AND DUBLIN.
1887.

CONTENTS.

ILLUSTRATIONS.

YUSSUF THE GUIDE.

CHAPTER I.

MEDICAL AND LEGAL.

BUT it seems so shocking, sir."

"Yes, madam," said the doctor, "very sad indeed. You had better get that prescription made up at once."

"And him drenched with physic!" cried Mrs. Dunn; "when it doesn't do him a bit of good."

"Not very complimentary to me, Mrs. Dunn," said the doctor smiling.

"Which I didn't mean any harm, sir; but wouldn't it be better to let the poor boy die in peace, instead of worrying him to keep on taking physic?"

"And what would you and his friends say if I did not prescribe for him?"

"I should say it was the best thing, sir; and as to his friends, why, he hasn't got any."

"Mr. Burne?"

"What! the lawyer, sir? I don't call him a friend. Looks after the money his poor pa left, and doles it

out once a month, and comes and takes snuff and
blows his nose all over the room, as if he was a human
trombone, and then says, 'Hum!' and 'Ha!' and 'Send
me word how he is now and then,' and goes away."

" But his father's executor, Professor Preston?"

" Lor' bless the man! don't talk about him. I wrote
to him last week about how bad the poor boy was;
and he came up from Oxford to see him, and sat down
and read something out of a roll of paper to him about
his dog."

"About his dog, Mrs. Dunn?"

"Yes, sir, about his dog Pompey, and then about
tombs—nice subject to bring up to a poor boy half-
dead with consumption! And as soon as he had done
reading he begins talking to him. You said Master
Lawrence was to be kept quiet, sir?"

" Certainly, Mrs. Dunn."

"Well, if he didn't stand there sawing one of his
hands about and talking — there, shouting — at the
poor lad as if he was in the next street, or he was a
hout-door preacher, till I couldn't bear it any longer,
and I made him go."

"Ah, I suppose the professor is accustomed to lecture."

" Then he had better go and lecture, sir. He sha'n't
talk my poor boy to death."

"Well, quiet is best for him, Mrs. Dunn," said the
doctor smiling at the rosy-faced old lady, who had
turned quite fierce; "but still, change and something
to interest him will do good."

" More good than physic, sir?"

"Well, yes, Mrs. Dunn, I will be frank with you—
more good than physic. What did Mr. Burne say

about the poor fellow going to Madeira or the south of France?"

"Said, sir, that he'd better take his Madeira out of a wine-glass and his south of France out of a book. I don't know what he meant, and when I asked him he only blew his nose till I felt as if I could have boxed his ears. But now, doctor, what do you really think about the poor dear? You see he's like my own boy. Didn't I nurse him when he was a baby, and didn't his poor mother beg of me to always look after him? And I have. Nobody can't say he ever had a shirt with a button off, or a hole in his clean stockings, or put on anything before it was aired till it was dry as a bone. But now tell me what you really think of him."

"That I can do nothing whatever, Mrs. Dunn," said the doctor kindly. "Our London winters are killing him, and I have no faith in the south of England doing any good. The only hope is a complete change to a warmer land."

"But I couldn't let him go to a horrible barbarous foreign country, sir."

"Not to save his life, Mrs. Dunn?"

"Oh, dear! oh, dear! oh, dear!" sighed the old lady. "It's very hard when I'd lay down my life to save him, and me seeing him peek and pine away and growing so weak. I know it was that skating accident as did it. Him nearly a quarter of an hour under the ice, and the receiving-house doctor working for an hour before he could bring him to."

"I'm afraid that was the start of his illness, Mrs. Dunn."

"I'm sure of it, doctor. Such a fine lad as he was, and he has never been the same since. What am I to do? Nobody takes any interest in the poor boy but me."

"Well, I should write at once to the professor and tell him that Mr. Lawrence is in a critical condition, and also to his father's executor, Mr. Burne, and insist upon my patient being taken for the winter to a milder clime."

"And they won't stir a peg. I believe they'll both be glad to hear that he is dead, for neither of them cares a straw about him, poor boy."

There had been a double knock while this conversation was going on in Guildford Street, Russell Square, and after the pattering of steps on the oil-cloth in the hall the door was opened, and the murmur of a gruff voice was followed by the closing of the front door, and then a series of three sounds, as if someone was beginning to learn a deep brass instrument, and Mrs. Dunn started up.

"It's Mr. Burne. Now, doctor, you tell him yourself."

Directly after, a keen-eyed gray little gentleman of about fifty was shown in, with a snuff-box in one hand, a yellow silk handkerchief in the other, and he looked sharply about as he shook hands in a hurried way, and then sat down.

"Hah! glad to see you, doctor. Now about this client of yours. Patient I mean. You're not going to let him slip through your fingers?"

"I'm sorry to say, Mr. Burne—"

"Bless me! I am surprised. Been so busy. Poor

boy! *Snuff snuff snuff.* Take a pinch? No, you said you didn't. Bad habit. Bless my soul, how sad!"

Mr. Burne, the family solicitor, jumped up when he blew his nose. Sat down to take some more snuff, and got up again to offer a pinch to the doctor.

"Really, Mr. Burne, there is only one thing that I can suggest—"

"And that's what Mrs. Dunn here told me."

There was a most extraordinary performance upon the nose, which made Mrs. Dunn raise her hands, and then bring them down heavily in her lap, and exclaim:

"Bless me, man, don't do that!"

"Ah, Mrs. Dunn," cried the lawyer; "what have you been about? Nothing to do but attend upon your young master, and you've got him into a state like this."

"Well of all—"

"Tut tut! hold your tongue, Mrs. Dunn, what's gone by can't be recalled. I've been very busy lately fighting a cousin of the poor boy, who was trying to get his money."

"And what's the good of his money, sir, if he isn't going to live?"

"Tut tut, Mrs. Dunn," said the lawyer, blowing his nose more softly, "but he is. I telegraphed to Oxford last night for Professor Preston to meet me here at eleven this morning. I have had no answer, but he may come. Eccentric man, Mrs. Dunn."

"Why you're never going to have him here to talk the poor boy to death."

"Indeed but I am, Mrs. Dunn, for I do not believe what you say is possible, unless done by a woman—

an old woman," said the lawyer looking at the old
lady fixedly.

"Well I'm sure!" exclaimed Mrs. Dunn, and the
doctor rose.

"You had better get that prescription made up,
Mrs. Dunn, and go on as before."

"One moment, doctor," said the lawyer, and he
drew him aside for a brief conversation to ensue.

"Bless me! very sad," said the lawyer; and then, as
Mrs. Dunn showed the doctor out, the old gentleman
took some more snuff, and then performed upon his
nose in one of the windows; opposite the fire; in one
corner; then in another; and then he was finishing
with a regular coach-horn blast when he stopped half-
way, and stared, for Mrs. Dunn was standing in the
doorway with her large florid cap tilted forward in
consequence of her having stuck her fingers in her ears.

"Could you hear me using my handkerchief, Mrs.
Dunn?" said the lawyer.

"Could I hear you? Man alive!" cried the old lady,
in a tone full of withering contempt, "could I hear
that?"

CHAPTER II.

THE SECOND GUARDIAN.

"THAT!" to which Mrs. Dunn alluded was a double knock at the front door; a few minutes later the maid ushered in a tall broad-shouldered man of about forty. His hair was thin upon the crown, but crisp and grizzled, and its spareness seemed due to the fact that nature required so much stuff to keep up the supply for his tremendous dark beard that his head ran short. It was one of those great beards that are supposed to go with the portrait of some old patriarch, and over this could be seen a pair of beautiful large clear eyes that wore a thoughtful dreamy aspect, and a broad high white forehead. He was rather shabbily dressed in a pepper-and-salt frock-coat, vest, and trousers, one of which had been turned up as if to keep it out of the mud while the other was turned down; and both were extremely baggy and worn about the knees. Judging from appearances his frock-coat might have been brushed the week before last, but it was doubtful, though his hat, which he placed upon the table as he entered, certainly had been brushed very lately, but the wrong way.

He did not wear gloves upon his hands, but in his trousers pockets, from which he pulled them to throw them in his hat, after he had carefully placed two great folio volumes, each minus one cover, upon a chair,

and then he shook hands, smiling blandly, with Mrs.
Dunn, and with the lawyer.

"Bless the man!" said Mrs. Dunn to herself, "one
feels as if one couldn't be cross with him; and there's
a button off the wrist-band of his shirt."

"'Fraid you had not received my telegram, sir," said
the lawyer in rather a contemptuous tone, for Mrs.
Dunn had annoyed him, and he wanted to wreak his
irritation upon someone else.

"Telegram?" said the professor dreamily. "Oh, yes.
It was forwarded to me from Oxford. I was in
town."

"Oh! In town?"

"Yes. At an hotel in Craven Street. I am making
preparations, you know, for my trip."

"No, I don't know," said the lawyer snappishly.
"How should I know?"

"Of course not," said the professor smiling. "The
fact is, I've been so much—among books—lately—
that—these are fine. Picked them up at a little shop
near the Strand. Buttknow's *Byzantine Empire.*"

He picked up the two musty old volumes, and
opened them upon the table, as a blast rang out.

The professor started and stared, his dreamy eyes
opening wider, but seeing that it was only the lawyer
blowing his nose, he smiled and turned over a few
leaves.

"A good deal damaged; but such a book is very
rare, sir."

"My dear sir, I asked you to come here to talk business," said the lawyer, tapping the table with his snuff-
box, "not books."

"True. I beg your pardon," said the professor. "I was in town making the final preparations for my departure to the Levant, and I did not receive the telegram till this morning. That made me so late."

"Humph!" ejaculated the lawyer, and he took some more snuff.

"And how is Lawrence this morning?" said the professor in his calm, mild way. "I hope better, Mrs. Dunn."

"Bless the man! No. He is worse," cried Mrs. Dunn shortly.

"Dear me! I am very sorry. Poor boy! I'm afraid I have neglected him. His poor father was so kind to me."

"Everybody has neglected him, sir," cried Mrs. Dunn, "and the doctor says that the poor boy will die."

"Mrs. Dunn, you shock me," cried the professor, with the tears in his eyes, and his whole manner changing. "Is it so bad as this?"

"Quite, sir," cried the lawyer, "and I want to consult you as my co-executor and trustee about getting the boy somewhere in the south of England or to France."

"But medical assistance," said the professor. "We must have the best skill in London."

"He has had it, sir," cried Mrs. Dunn, "and they can't do anything for him. He's in a decline."

"There, sir, you hear," said the lawyer. "Now, then, what's to be done?"

"Done!" cried the professor, with a display of animation that surprised the others. "He must be removed to a warmer country at once. I had no idea that matters were so bad as this. Mr. Burne, Mrs. Dunn, I am a student much interested in a work I am writing on the Byzantine empire, and I was starting in a few days for Asia Minor. My passage was taken. But all that must be set aside, and I will stop and see to my dear old friend's son."

Poo woomp poomp. Pah!

Mr. Burne blew a perfectly triumphal blast with his pocket-handkerchief, took out his snuff-box, put it back, jumped up, and, crossing to where the professor was standing, shook his hand very warmly, and without a word, while Mrs. Dunn wiped her eyes upon her very stiff watered silk apron, but found the result so unsatisfactory that she smoothed it down, and hunted out a pocket-handkerchief from somewhere among the folds of her dress and polished her eyes dry.

Then she seemed as if she put a sob in that piece of white cambric, and wrapped it up carefully, just as if it were something solid, doubling the handkerchief over and over and putting it in her pocket before going up to the professor and kissing his hand.

"Ha!" said the latter, smiling at first one and then the other. "This is very good of you. I don't often find people treat me so kindly as this. You see, I am such an abstracted, dreamy man. I devote myself so much to my studies that I think of nothing else. My friends have given me up, and—and I'm afraid they laugh at me. I am writing, you see, a great work upon the old Roman occupation of—. Dear me! I'm wandering off again. Mrs. Dunn, can I not see my old friend's son?"

(348)

"To be sure you can, sir. Pray, come," cried the old lady; and, leading the way, she ushered the two visitors out into the hall, the professor following last, consequent upon having gone back to fetch the two big folio volumes; but recollecting himself, and colouring like an ingenuous girl, he took them back, and laid them upon the dining-room table.

Mrs. Dunn paused at the drawing-room door and held up a finger.

"Please, be very quiet with him, gentlemen," she said. "The poor boy is very weak, and you must not stay long."

The lawyer nodded shortly, the professor bent his head in acquiescence, and the old lady opened the drawing-room door.

CHAPTER III.

A PLAN IS MADE.

S they entered, a pale attenuated lad of about seventeen, who was lying back in an easy chair, with his head supported by a pillow, and a book in his hand, turned to them slightly, and his unnaturally large eyes had in them rather a wondering look, which was succeeded by a smile as the professor strode to his side, and took his long, thin, girlish hand.

"Why, Lawrence, my boy, I did not know you were so ill."

" Ill? Nonsense, man!" said the lawyer shortly.
" He's not ill. Are you, my lad?"

He shook hands rather roughly as he spoke from
the other side of the invalid lad's chair, while Mrs.
Dunn gave her hands an impatient jerk, and went behind to brush the long dark hair from the boy's forehead.

He turned up his eyes to her to smile his thanks,
and then laid his cheek against the hand that had been
smoothing his hair.

" No, Mr. Burne, I don't think I'm ill," he said in a
low voice. " I only feel as if I were so terribly weak
and tired. I get too tired to read sometimes, and I
never do anything at all to make me so."

" Hah!" ejaculated the lawyer.

" I thought it was the doctor come back," continued
the lad. " I say, Mr. Preston—you are my guardian,
you know—is there any need for him to come? I am
so tired of cod-liver oil."

" Yah!" ejaculated the lawyer; " it would tire anybody but a lamp."

He snorted this out, and then blew another blast
upon his nose, which made some ornament upon the
chimney-piece rattle.

" Doctor?" said the professor rather dreamily, as he
sat down beside the patient. " I suppose he knows
best. I did not know you were so ill, my boy."

" I'm not ill, sir."

" But they say you are, my lad. I was going abroad,
but I heard that you were not so well, and—and I
came up."

" I am very glad," said the lad, " for it is very dull

lying here. Old Dunny is very good to me, only she will bother me so to take more medicine, and things that she says will do me good, and I do get so tired of everything. How is the book getting on, sir?"

"Oh, very slowly, my lad," said the professor, with more animation. "I was going abroad to travel and study the places about which I am writing, but—"

"When do you go?" cried the lad eagerly.

"I was going within a few days, but—"

"Where to?"

"Smyrna first, and then to the south coast of Asia Minor, and from thence up into the mountains."

"Is it a beautiful country, Mr. Preston?"

"Yes; a very wild and lovely country, I believe."

"With mountains and valleys and flowers?"

"Oh, yes, a glorious place."

"And when are you going?"

"I was going within a few days, my boy," said the professor kindly; "but—"

"Is it warm and sunshiny there, sir?"

"Very."

"In winter?"

"Oh, yes, in the valleys; in the mountains there is eternal snow."

"But it is warm in the winter?"

"Oh, yes; the climate is glorious, my lad."

"And here, before long, the leaves will fall from that plane-tree in the corner of the square, that one whose top you can just see; and it will get colder, and the nights long, and the gas always burning in the lamps, and shining dimly through the blinds; and then the fog will fill the streets, and creep in through the

cracks of the window; and the blacks will fall and
come in upon my book, and it will be so bitterly
cold, and that dreadful cough will begin again. Oh,
dear!"

There was silence in the room as the lad finished
with a weary sigh; and though it was a bright
morning in September, each of the elder personages
seemed to conjure up the scenes the invalid portrayed,
and thought of him lying back there in the desolate
London winter, miserable in spirit, and ill at ease from
his complaint.

Then three of the four present started, for the lawyer
blew a challenge on his trumpet.

"There is no better climate anywhere, sir," he said,
addressing the professor, "and no more healthy spot
than London."

"Bless the man!" ejaculated Mrs. Dunn.

"I beg to differ from you, sir," said the professor in
a loud voice, as if he were addressing a class. "By
the reports of the meteorological society—"

"Hang the meteorological society, sir!" cried the
lawyer, "I go by my own knowledge."

"Pray, gentlemen!" cried Mrs. Dunn, "you forget
how weak the patient is."

"Hush, Mrs. Dunn," said the lad eagerly; "let them
talk. I like to hear."

"I beg pardon," said the professor; "and we are
forgetting the object of our visit. Lawrence, my boy,
would you like to go to Brighton or Hastings, or the
Isle of Wight?"

"No," said the lad sadly, "it is too much bother."

"To Devonshire, then—to Torquay?"

"No, sir. I went there last winter, and I believe it made me worse. I don't want to be always seeing sick people in invalid chairs, and be always hearing them talk about their doctors. How long shall you be gone, sir?"

"How long? I don't know, my lad. Why?"

The boy was silent, and lay back gazing out of the window in a dreamy way for some moments before he spoke again, and then his hearers were startled by his words.

"I feel," he said, speaking as if to himself, "as if I should soon get better if I could go to a land where the sun shone, and the sea was blue, and the sweet soft cool breezes blew down from the mountains that tower up into the clear sky—where there were fresh things to see, and there would be none of this dreadful winter fog."

The professor and the lawyer exchanged glances, and the latter took a great pinch of snuff out of his box, and held it half-way up towards his nose.

Then he started, and let it fall upon the carpet— so much brown dust, for the boy suddenly changed his tone, and in a quick excited manner exclaimed, as he started forward:

"Oh! Mr. Preston, pray—pray—take me with you when you go."

"But, my dear boy," faltered the professor, "I am not going now. I have altered my plans."

"Then I must stop here," cried the boy in a passionate wailing tone—stop here and die."

There was a dead silence once more as the lad covered his face with his thin hands, only broken by

Mrs. Dunn's sobs as she laid her head upon the back
of the chair and wept aloud, while directly after Mr.
Burne took out his yellow handkerchief, prepared for
a blow, and finally delivered himself of a mild and
gentle sniff.

"Lawrence!"

It was the deep low utterance of a strong man who
was deeply moved, and as the boy let fall his thin
white fingers from before his eyes he saw that the professor was kneeling by his chair ready to take one of
his hands and hold it between his broad palms.

"Lawrence, my boy," he said; "your poor father and
I were great friends, and he was to me as a brother;
your mother as a sister. He left me as it were the
care and charge of you, and it seems to me that in my
selfish studies I have neglected my trust; but, Heaven
helping me, my boy, I will try and make up for the past.
You shall go with me, my dear lad, and we will search
till we find a place that shall restore you to health and
strength."

"You will take me with you?" cried the boy with a
joyous light in his eyes.

"That I will," cried the professor.

"And when?"

"As soon as you can be moved."

"But," sighed the lad wearily, "it will cost so
much."

"Well?" said the professor, "what of that? I am
not a poor man. I never spend my money."

"Oh! if it came to that," said the lawyer, taking
some more snuff and snapping his fingers, "young
Lawrence here has a pretty good balance lying idle."

"Mr. Burne, for shame!" cried Mrs. Dunn; "here have I been waiting to hear you speak, and you encourage the wild idea, instead of stamping upon it like a black beadle."

"Wild idea, ma'am?" cried the lawyer, blowing a defiant blast.

"Yes, sir; to talk about taking that poor weak sickly boy off into foreign lands among savages, and cannibals, and wild beasts, and noxious reptiles."

"Stuff, ma'am, stuff!"

"But it isn't stuff, sir. The doctor said—"

"Hang the doctor, ma'am!" cried the lawyer. "The doctor can't cure him, poor lad, so let's see if we can't do a little better."

"Why, I believe you approve of it, sir!" cried Mrs. Dunn with a horror-stricken look.

"Approve of it, ma'am? To be sure, I do. The very thing. Asia Minor, didn't you say, Mr. Preston?"

The professor bowed.

"Yes; I've heard that you get summer weather there in winter. I think you have hit the right nail on the head."

"And you approve of it, sir?" cried the boy excitedly.

"To be sure, I do, my lad."

"It will kill him," said Mrs. Dunn emphatically.

"Tchah! stuff and nonsense, ma'am!" cried the lawyer. "The boy's too young and tough to kill. We'll take him out there and make a man of him."

"We, sir?" exclaimed the professor.

"Yes, sir, we," said the lawyer, taking some more snuff, and dusting his black waistcoat. "Hang it all!

Do you think you are the only man in England who wants a holiday?"

"I beg your pardon," said the professor mildly; "of course not."

"I haven't had one worth speaking of," continued Mr. Burne, "for nearly—no, quite thirty years, and all that time I've been in dingy stuffy Sergeant's Inn, sir. Yes; we'll go travelling, professor, and bring him back a man."

"It will kill him," cried Mrs. Dunn fiercely, and ruffling up and coming forward like an angry hen in defence of her solitary chick, the last the rats had left.

The lawyer sounded his trumpet, as if summoning his forces to a charge.

"I say he shall not go."

"Mrs. Dunn," began the professor blandly.

"Stop!" cried the lawyer; "send for Doctor Shorter."

"But he has been, sir," remonstrated Mrs. Dunn.

"Then let him come again, ma'am. He shall have his fee," cried the lawyer; "send at once."

Mrs. Dunn's lips parted to utter a protest, but the lawyer literally drove her from the room, and then turned back, taking snuff outrageously, to where the professor was now seated beside the sick lad.

"That's routing the enemy," cried the lawyer fiercely. "Why, confound the woman! She told me that the doctor said he ought to be taken to a milder clime."

"But do you really mean, Mr. Burne, that, supposing the doctor gives his consent, you would accompany us abroad?"

"To be sure I do, sir, and I mean to make myself as unpleasant as I can. I've a right to do so, haven't I?"

"Of course," said the professor coldly.

"And I've a right to make myself jolly if I like, haven't I, sir?"

"Certainly," replied the professor, gazing intently at the fierce grizzled little man before him, and wondering how much he spent a-year in snuff.

"It will not cost you anything, and I shall not charge my expenses to the estate, any more than I shall let you charge yours, sir."

"Of course not, sir," said the professor more coldly still, and beginning to frown.

"You shall pay your expenses, I'll pay mine, and young Lawrence here shall pay his; and I tell you what, sir, we three will have a thoroughly good outing. We'll take it easy, and we'll travel just where you like, and while you make notes, Lawrence here and I will fish and run about and catch butterflies, eh? Hang it, I haven't caught a butterfly these three or four and thirty years, and I think it's time I had a try. Eh, what are you laughing at, sir?"

Lawrence Grange's laugh was low and feeble, but it brightened up his sad face, and was contagious, for it made the professor smile as well. The cold stern look passed away, and he held out his hand to the lawyer.

"Agreed, sir," he said. "If the doctor gives his consent, we will all three go, and, please Heaven, we will restore our young friend here his health and strength."

"Agreed, sir; with the doctor's consent or without," cried the lawyer, grasping the extended hand. "By George, we must begin to make our preparations at once! and as for the doctor— Oh, here he is!"

For there was a double knock, and directly after

Mrs. Dunn, appearing very much agitated, ushered in the doctor, who did not look quite so cool as he did when he left.

"Oh!" he ejaculated, "I was afraid from Mrs. Dunn's manner that something was wrong."

"No, doctor, nothing," said the lawyer. "We only want to ask you what you think of our young friend here being taken to spend the winter in Turkey."

"Admirable!" said the doctor, "if it could be managed."

"Oh, Doctor Shorter!" wailed Mrs. Dunn, "I thought you would stop this mad plan."

"There, madam, there!" cried the lawyer; "what did I say?"

"But he is not fit to move," cried Mrs. Dunn, while the boy's cheeks were flushed, and his eyes wandered eagerly from speaker to speaker.

"Only with care," said the doctor. "I should not take a long sea trip, I think; but cross to Paris, and then go on gently, stopping where you pleased, to Brindisi, whence the voyage would be short."

"The very thing!" cried the lawyer, giving one emphatic blow with his nose. "What do you say, professor?"

"It is the plan I had arranged if I had gone alone," was the reply; "and I think if Doctor Shorter will furnish us with the necessary medicines—"

"He requires change more than medicines," said the doctor. "Care against exertion, and—there, your own common sense will tell you what to do."

"Doctor! doctor! doctor!" sobbed Mrs. Dunn; "I didn't think it of you. What's to become of me?"

"You, madam?" replied the doctor. "You can read and write letters to our young friend here, and thank Heaven that he has friends who will take him in charge and relieve him from the risk of another winter in our terrible climate."

"Hear, hear!" and "No, no!" cried the lawyer. "Doctor Shorter, ours is not a bad climate, and I will not stand here and listen to a word against it. Look at me, sir! Thirty years in Sergeant's Inn—fog, rain, snow, and no sunshine; and look at me, sir—look at me!"

"My dear sir," said the doctor smiling, "you know the old saying about one man's meat being another man's poison? Suppose I modify my remark, and say terrible climate for our young friend. You are decided, then, to take him?"

"Certainly," said the professor.

"To Turkey?"

"Turkey in Asia, sir, where I propose to examine the wonderful ruins of the ancient Greek and Roman cities."

"And hunt up treasures of all kinds, eh?" said the doctor smiling.

"I hope we may be fortunate enough to discover something worthy of the search."

"But, let me see—the climate; great heat in the plains; intense cold in the mountains; fever and other dangers. You must be careful, gentlemen. Brigands —real brigands of the fiercest kind—men who mean heavy ransoms, or chopped-off heads. Then you will have obstinate Turks, insidious and tricking Greeks, difficulties of travel. No child's play, gentlemen."

"The more interest, sir," replied the professor, "the greater change."

"Well," said the doctor, "I shall drop in every day till you start, and be able to report upon our friend's health. Now, good day."

The doctor left the room with Mrs. Dunn, and as he went out Mr. Burne blew a flourish, loud enough to astonish the professor, who wondered how it was that so much noise could be made by such a little man, till he remembered the penetrating nature of the sounds produced by such tiny creatures as crickets, and then he ceased to be surprised.

CHAPTER IV.

A VERBAL SKIRMISH.

IT seemed wonderful: one day in London, then the luggage all ticketed, the young invalid carefully carried by a couple of porters to a first-class carriage, and seated in a snug corner, when one of them touched his cap and exclaimed:

"Glad to see you come back, sir, strong enough to carry me. Pore young chap!" he said to his mate; "it do seem hard at his time o' life."

"Hang the fellow!" cried the lawyer; "so it does at any time of life. I don't want to be carried by a couple of porters."

Then there was a quick run down to Folkestone,

with the patient tenderly watched by his two companions, the professor looking less eccentric in costume, for he had trusted to his tailor to make him some suitable clothing; but the lawyer looking more so, for he had insisted upon retaining his everyday-life black frock-coat and check trousers, the only change he had made being the adoption of a large leghorn straw-hat with a black ribbon; on the whole as unsuitable a costume as he could have adopted for so long a journey.

"But I've got a couple of Holland blouses in one of my portmanteaus," he said to Lawrence, "and these I shall wear when we get into a hotter country."

At Folkestone, Lawrence showed no fatigue; on the contrary, when the professor suggested staying there for the night he looked disappointed, and begged that they might cross to Boulogne, as he was so anxious to see France.

Judging that it was as well not to disappoint him, and certainly advisable to take advantage of a lovely day with a pleasant breeze for the crossing, the professor decided to proceed—after a short conversation between the two elders, when a little distant feeling was removed, for the professor had felt that the lawyer was not going to turn out a very pleasant travelling companion.

"What do you think, sir?" he had said to the fierce-looking little man, who kept on attracting attention by violently blowing his nose.

"I'll tell you what I think, professor," was the reply. "It seems to me that the boy is a little sore and upset with his parting from his old nurse. Milksoppish, but natural to one in his state. He wants to get right away, so as to forget the trouble in new impressions. Then, as you see, the journey so far has not hurt him, and he feels well enough to go on. Sign, sir, that nature says he is strong enough, so don't thwart him. Seems to me, sir—*snuff, snuff, snuff*—that the way to do him good is to let him have his own way, so long as he doesn't want to do anything silly. Forward!"

So they went forward, a couple of the steamer's men lifting Lawrence carefully along the gangway and settling him in a comfortable part of the deck, which he preferred to going below; and ten minutes later the machinery made the boat quiver, the pier seemed to be running away, and the professor said quietly: "Good-bye to England."

The sea proved to be more rough than it had seemed from the pier, and, out of about seventy passengers, it was not long before quite sixty had gone below, leaving the deck very clear; and the professor, who kept walking up and down, while the lawyer occupied a seat near Lawrence, kept watching the invalid narrowly.

But there was no sign of illness. The lad looked terribly weak and delicate, but his eyes were bright, and the red spots on his cheeks were unchanged.

"I say, Preston," said the lawyer, when they had been to sea about a quarter of an hour, "you look very pale; if you'd like to go below I'll stay with him."

"Thanks, no," was the reply; "I prefer the deck. How beautiful the chalky coast looks, Lawrence!"

"Yes, lovely," was the reply; "but I was trying to

look forward to see France. I want to see health. Looking back seems like being ill."

The professor nodded, and said that the French coast would soon be very plain, and he stalked up and down, a magnificent specimen of humanity, with his great beard blown about by the wind, which sought in vain to play with his closely-cut hair.

"I'm sure you had better go below, professor. You look quite white," said the lawyer again; but Mr. Preston laughed.

"I am quite well," he said; and he took another turn up and down to look at the silvery foam churned up by the beating paddles.

"Look here!" cried the lawyer again, as the professor came and stood talking to Lawrence; "had you not better go down?"

"No. Why go down to a cabin full of sick people, when I am enjoying the fresh air, and am quite well?"

"But are you really quite well?"

"Never better in my life."

"Then it's too bad, sir," cried the lawyer. "I've been waiting to see you give up, and if you will not, I must, for there's something wrong with this boat."

"Nonsense! One of the best boats on the line."

"Then, there's something wrong with me. I can't enjoy my snuff, and it's all nonsense for this boy to be called an invalid. I'm the invalid, sir, and I am horribly ill. Help me below, there's a good fellow."

Mr. Burne looked so deplorably miserable, and at the same time so comic, that it was impossible to avoid smiling, and as he saw this he stamped his foot.

"Laughing at me, eh? Both of you. Now, look

here. I know you both feel so poorly that you don't know what to do, and I'll stop up on deck and watch you out of spite."

"Nonsense! I could not help smiling," said the professor good-humouredly. "Let me help you down."

"Thank you, no," said the lawyer taking off his hat to wipe his moist brow, and then putting it on again, wrong way first. "I'm going to stop on deck, sir—to stop on deck."

He seemed to be making a tremendous effort to master the qualmish feeling that had attacked him, and in this case determination won.

A night at Boulogne, and at breakfast-time next morning Lawrence seemed no worse for the journey, so they went on at once to Paris, where a day's rest was considered advisable, and then, the preliminaries having been arranged, the train was entered once more, and after two or three stoppages to avoid over-wearying the patient, Trieste was reached, where a couple of days had to be passed before the arrival of the steamer which was to take them to Smyrna, and perhaps farther, though the professor was of opinion that it might be wise to make that the starting-place for the interior.

But when the steamer arrived a delay of five days more ensued before a start was made; and all this time the invalid's companions watched him anxiously.

It was in these early days a difficult thing to decide, and several times over the professor and Mr. Burne nearly came to an open rupture—one sufficiently serious to spoil the prospects of future friendly feeling.

But these little tiffs always took place unknown to Lawrence, who remained in happy ignorance of what was going on.

The disagreements generally happened something after this fashion.

Lawrence would be seated in one of the verandahs of the hotel enjoying the soft warm sea-breeze, and gazing out at the scene glowing in all the brightness of a southern sun, when the old lawyer would approach the table where, out of the lad's sight and hearing, the professor was seated writing.

The first notice the latter had of his fellow-traveller's approach would be the loud snapping of the snuff-box, which was invariably followed by a loud snuffling noise, and perhaps by a stentorian blast. Then the lawyer would lean his hand upon the table where the professor was writing with:

"Really, my dear sir, you might put away your pens and ink for a bit. I've left mine behind. Here, I want to talk to you."

The professor politely put down his pen, leaned back in his chair and folded his arms.

"Hah! that's better," said Mr. Burne. "Now we can talk. I wanted to speak to you about that boy."

"I am all attention," said the professor.

"Well, sir, there's a good German physician here as well as the English one. Don't you think we ought to call both in, and let them have a consultation?"

"What about?" said the professor calmly.

"About, sir? Why, re Lawrence."

"But he seems certainly better, and we have Doctor Shorter's remedies if anything is necessary."

"Better, sir? decidedly worse. I have been watching him this morning, and he is distinctly more feeble."

"Why, my dear Mr. Burne, he took my arm half an hour ago, and walked up and down that verandah without seeming in the least distressed."

"Absurd, sir!"

"But I assure you—"

"Tut, tut, sir! don't tell me. I watch that boy as I would an important case in a court of law. Nothing escapes me, and I say he is much worse."

"Really, I should be sorry to contradict you, Mr. Burne," replied the professor calmly; "but to me it seems as if this air agreed with him, and I should have said that, short as the time has been since he left home, he is better."

"Worse, sir, worse decidedly."

"Really, Mr. Burne, I am sorry to differ from you," replied the professor stiffly; "but I must say that Lawrence is, to my way of thinking, decidedly improved."

"Pah! Tchah! Absurd!" cried the lawyer; and he went off blowing his nose.

Another day he met the professor, who had just left Lawrence's side after sitting and talking with him for some time, and there was an anxious, care-worn look in his eyes that impressed the sharp lawyer at once.

"Hallo!" he exclaimed; "what's the matter?"

The professor shook his head.

"Lawrence," he said sadly.

"Eh? Bless me! You don't say so," cried Mr. Burne; and he hurried out into the verandah, which was the lad's favourite place.

There Mr. Burne stayed for about a quarter of an hour, and then went straight to where the professor was writing a low-spirited letter to Mrs. Dunn, in which he had said that he regretted bringing Lawrence right away into those distant regions, for though Trieste was a large port, and there was plenty of medical attendance to be obtained, it was not like · being at home.

"I say! Look here!" cried Mr. Burne, "you ought to know better, you know."

"I do not understand you," replied the professor quietly.

"Crying wolf, you know. It's too bad."

"Really," said the professor, who was in one of his dreamy, abstracted moods, "you are mistaken, Mr. Burne. I did not say a word about a wolf."

"Well, whoever said you did, man?" cried the lawyer impatiently as he took out his snuff-box and whisked forth a pinch, flourishing some of the fine dry dust about where he stood. "Can't you, a university man, understand metaphors — shepherd boy calling wolf when there was nothing the matter? The patient's decidedly better, sir."

"Really, Mr. Burne—er—tchishew—er—tchishew!"

Old Mr. Burne stood looking on, smiling grimly, as the professor had a violent fit of sneezing, and in mocking tones held out his snuff-box and said:

"Have a good pinch? Stop the sneezing. Ah! that's better," he added, as the professor finished off with a tremendous burst. "Your head will be clear now, and you can understand what I say. That boy's getting well."

" I wish I could think so," said the professor, sniffing so very quietly that, as if to give him a lesson, his companion blew off one of his blasts, with the result that a waiter hurried into the room to see what was wrong.

" Think? there is no occasion to think so. He is mending fast, sir; and if you have any doubt about it, and cannot trust in the opinion of a man of the world, go and watch him, and see how interested he seems in all that is going on. Why, a fortnight ago he lay back in his chair dreaming and thinking of nothing but himself. Now he is beginning to forget that there is such a person. He's better, sir, better."

The fact was that the lawyer was right, and so was the professor, for at that time Lawrence was as changeable of aspect as an April day, and his friends could only judge him by that which he wore when they went to his side.

At last the morning came when the steamer started for Smyrna, and the pair were for once in a way agreed. They had been breakfasting with Lawrence, noting his looks, his appetite, listening to every word, and at last, when he rose feebly, and went out into the verandah to gaze down at the busy crowd of mingled European and Eastern people, whose dress and habits seemed never tiring to the lad, the lawyer turned to the professor and exclaimed:

" You did not say a word to him about sailing to-day."

" No. Neither did you."

" Well, why didn't you?"

" Because I thought that it seemed useless, and that we had better stay."

"Well, I don't often agree with you, professor, but I must say that I do to-day. The boy is not equal to it. But he is better."

"Ye—es," said the professor. "I think he is better."

Just then Lawrence returned from the verandah, looking flushed and excited.

"Why, the Smyrna boat sails to-day, Mr. Preston," he exclaimed. "One of the waiters has just told me. Hadn't we better get ready at once?"

"Get ready?" said the professor kindly. "We thought that perhaps we had better wait for the next boat."

"Oh!" exclaimed Lawrence, with his countenance changing. "I shall be so disappointed. I felt so much better too, and I've been longing to see some of the Grecian isles."

"Do you really feel yourself equal to the journey, my dear boy?" said the professor.

"Oh yes. I don't know when I have felt so well," said Lawrence eagerly.

"Bless my soul!" cried the old lawyer, opening and shutting his snuff-box as if for the purpose of hearing it snap, and sending the fine dust flying, "what a young impostor you are! Here, let's get our bill paid, and our traps on board. There's no time to spare."

Lawrence's face brightened again, and he left the room.

"Tell you what, professor," said Mr. Burne, "you and I have been ready to quarrel several times over about what we do not understand. Now, look here. I want to enjoy this trip. What do you say to burying the hatchet?"

"Burying the hatchet? Oh! I see. Let there be peace."

"To be sure," cried the lawyer, shaking hands warmly, "and we'll keep the fighting for all the Greeks, Turks, brigands, and the like who interfere with us."

"With all my heart," said the professor smiling; but Mr. Burne still lingered as if he had something to say.

"Fact is," he exclaimed at last, "I'm a curious crotchety sort of fellow. Had too much law, and got coated over with it; but I'm not bad inside when you come to know me."

"I'm sure you are not, Burne," said the professor warmly; "and if you come to that, I have spent so many years dealing with dead authors, and digging up musty legends, that I am abstracted and dreamy. I do not understand my fellow-men as I should, but really I esteem you very highly for the deep interest you take in Lawrence."

"That's why I esteem you, sir," said the lawyer; "and — no, I won't take any more snuff now; it makes you sneeze. There, be off, and get ready while I pay the bills."

That evening, in the golden glow of the setting sun, they set sail for Smyrna.

CHAPTER V.

SOME FELLOW-TRAVELLERS.

IT was one bright morning, after a delightful passage, that the steamer made its way into the port of Smyrna, where everything around seemed to be full of novelty— strange craft manned by strange-looking crews, Turks with white turbans, Turks with scarlet fezzes and baggy breeches, and Turks with green turbans to show their reputation among their compatriots. Greeks, too—small, lithe, dark men, with keen faces and dark eyes, differing wonderfully from the calm, dignified, handsome Turks, but handsome in their way if it had not been for a peculiarly sharp, shifty expression that suggested craftiness and a desire to overreach, if not cheat.

There was a constant succession of fresh sights, from the Turkish man-of-war that was of British build, to the low fishing-boat with its long graceful lateen sail, spread out upon its curved and tapering spar.

Ashore it was the same. The landing-place swarmed with fresh faces, fresh scenes. Everything looked bright, and as if the atmosphere was peculiarly clear, while the shadows were darker and sharper as they were cast by the glowing sun.

For the sun did glow. The time was short since they had left England, with symptoms coming on of falling leaves, lengthening nights, and chills in the air, while here all was hot summer time, and one of the first things Mr. Burne said was:

"There's no mistake about it, I must have out a blouse."

They were soon comfortably settled in the best hotel, from whence the professor decided to sally forth at once to call upon and deliver his letters of recommendation to the British consul; but he was not fated to go alone.

"I want to see everything and everybody," said Mr. Burne, "and I'll go with you. Look here, Lawrence, my boy, I would not get in the sun. I'd go and lie down for an hour or two till we get back."

"The sun seems to give me strength," said Lawrence eagerly. "I have seen so little of it in London. I want to go with you, please."

The professor darted a look at Mr. Burne which seemed to say, "Let him have his own way;" and the landlord having been consulted, a Greek guide or dragoman was soon in readiness, and they started.

"Look here," said Mr. Burne, taking hold of the professor's sleeve. "I don't like the look of that chap."

"What, the guide?"

"Yes! I thought Greeks were nice straightforward chaps, with long noses drawn down in a line from their foreheads, like you see in the British Museum. That fellow looks as if he wouldn't be long in England before he'd be looking at a judge and jury, and then be sent off to penal servitude. Greek statues are humbug. They don't do the Greeks justice."

"It does not matter as long as he does his duty by us

for the short time we are here. Be careful. He understands English."

"Well, I am careful," said Mr. Burne; "and I'm looking after my pocket-book, watch, and purse; and if I were you I should do the same. He's a rogue, I'm sure."

"Nonsense!"

"'Tisn't nonsense, sir; you're too ready to trust everybody. Did you hear his name?"

"I did," said Lawrence smiling. "Xenos Stephanos."

"Yes," grumbled Mr. Burne. "There's a name. I don't believe any man could be honest with a name like that."

The professor showed his white teeth as he laughed heartily, and Mr. Burne took snuff, pulled out a glaring yellow silk handkerchief, and blew a blast that was like the snort of a wild horse.

It was done so suddenly that a grave-looking Turkish gentlemen in front started and turned round.

"Well, what is it?" said Mr. Burne fiercely. "Did you never see an Englishman take snuff before?"

The Turk bowed, smiled, and continued his way.

"Such rudeness. Savages!" snorted out Mr. Burne. "Don't believe they know what a pocket-handkerchief is."

"I beg your pardon," said the Turk, turning round and smiling as he spoke in excellent English, "I think you will find we do, but we have not the use for them here that you have in England."

"I—er—er—er. Bless my soul, sir! I beg your pardon," cried the old lawyer. "I did not know you understood English, or—"

"Pray, say no more, sir," said the Turkish gentleman gravely. And he turned to cross the street.

"Snubbed! Deserved it!" cried Mr. Burne, taking off his straw hat, and doubling his fist, as if he were going to knock the crown out. "Let this be a lesson to you, Lawrence. Bless me! Thought I was among savages. Time I travelled."

"You forgot that you were still amongst steam, and post-offices, and telegraph wires, and—"

"Bless me! yes," cried Mr. Burne; "and, look there, an English name up, and Bass's pale ale. Astonishing!"

Just then the Greek guide stopped and pointed to a private house as being the English consul's, and upon entering they were at once shown into a charmingly furnished room, in which were a handsome bronzed middle-aged gentleman, in earnest conversation with a tall masculine-looking lady with some pretensions to beauty, and a little easy-looking man in white flannel, a glass in one eye, and a very high shirt collar covered with red spots, as if a number of cochineal insects had been placed all over it at stated intervals and then killed.

He was smooth-faced all but a small moustache; apparently about thirty; plump and not ill-favoured, though his hair was cut horribly close; but a spectator seemed to have his attention taken up at once by the spotted collar and the eye-glass.

"Glad to see you, Mr. Preston," said the bronzed middle-aged man. "You too, Mr. Burne. And how are you, Mr. Grange? I hope you have borne the voyage well. Let me introduce you," he continued, after shaking hands, "to our compatriots Mr. and Mrs. Charles Chumley. We can't afford, out here, not to know each other."

Mutual bowing took place, and the consul continued:

"Mr. and Mrs. Chumley are bound on the same errand as you are—a trip through the country here."

"Yes," said the gentleman; "we thought—"

"Hush, Charley! don't," interrupted the lady; "let me speak. Are you Professor Preston?"

"My name is Preston," said the professor, bowing.

"Glad to meet you. Mr. Chumley and I are going to do Turkey this year. Mr. Thompson here said that you and your party were going to travel. He had had letters of advice. We are going to start directly and go through the mountains; I suppose you will do the same."

"No," said the professor calmly; "we are going to take steamer round to one of the southern ports and start from there."

"Oh, I say, what a pity!" said the little gentleman, rolling his head about in his stiff collar, where it looked something like a ball in a cup. "We might have helped one another and been company."

"I wish you would not interfere so, Charley," cried the lady. "You know what I said."

"All right, Agnes," said the little gentleman dolefully. "Are you people staying at Morris's?"

"Yes," said the professor.

"So are we. See you at dinner, perhaps."

"Charley!" exclaimed the lady in tones that were quite Amazonic, they were so deep and stern.

Then a short conversation took place with the consul, and the strange couple left, leaving their host free to talk to the other visitors.

"I had very kind letters from Mr. Linton at the Foreign Office respecting you, gentlemen," said the consul.

"I know Linton well," said the professor.

"He is an old friend of mine too," said the consul. "Well, I have done all I could for you."

"About passports or what is necessary?" said the professor.

"I have a properly-signed firman for you," said the consul smiling; "and the showing of that will be sufficient to ensure you good treatment, help, and protection from the officials in every town. They will provide you with zaptiehs or cavasses—a guard when necessary, and generally see that you are not molested or carried off by brigands, or such kind of folk."

"But is it a fact, sir," said Mr. Burne, "that you have real brigands in the country?"

"Certainly," said the consul smiling.

"What! in connection with postal arrangements, and steam, and telegraphs?"

"My dear sir, we have all these things here; but a score or so of miles out in the country, and you will find the people, save that firearms are common, just about as they were a thousand years ago."

"Bless my heart!" exclaimed Mr. Burne.

"It is a fact, sir; and I should advise great care, not only as to whom you trust among the people, but as to your health. The country is in a horrible state of neglect; the government does nothing."

"But I do not see how that is to affect us," said the professor, "especially as we have that firman."

"It will not affect you in the more settled districts, but you may run risks in those which are more remote. I have been warning Mr. and Mrs. Chumley about the risks, but the lady laughed and said that she always carried a revolver."

"Bless me!" exclaimed Mr. Burne, "a lady with a revolver! She would not dare to fire it."

"I don't know about that," said the professor.

"Of course," continued the consul, "I am at your service, Mr. Preston. If you are in need of aid, and are anywhere within reach of the telegraph wires, pray send to me and I will do my best. Can I do anything more for you?"

This was a plain hint to go, for it was evident that others were waiting for an interview with the representative of England; so a friendly farewell was taken and the little party returned to the hotel.

"I'm glad you decided to go a different way to those people, Preston," said Mr. Burne.

"The decision was made on the instant, my dear sir; for I did mean to start from here."

"Ah, you thought those people would be a nuisance?"

"Indeed I did."

The professor had hardly spoken when Lawrence touched his arm; for the parties alluded to approached, and the lady checked her lord, who was going to speak, by saying:

"I thought I would give you a hint about going pretty well armed. You will not have to use your weapons if you let the people see that you have them."

"Arms, ma'am! Stuff! rubbish!" cried Mr. Burne. "The proper arms of an Englishman are the statutes at large, bound in law calf, with red labels on their back."

"Statutes at large!" said the lady wonderingly.

"Yes, ma'am—the laws of his country, or the laws of the country where he is; and the proper arms of a lady, madam, are her eyes."

"And her tongue," said the professor to himself, but not in so low a voice that it was not heard by Lawrence, who gave him a sharp look full of amusement.

Mrs. Chumley smiled and bowed.

"Very pretty, sir!" she said; "but you forget that we are going to travel through a country where the laws are often a mere name, and people must take care of themselves."

"Take care of themselves—certainly, ma'am, but not by breaking the laws. If a pack of vagabonds were to attack me I should hand them over to the police, or apply at the nearest police-court for a summons. That would be a just and equitable way of treating the matter."

"Where would you get your police, Burne? and whom would you get to serve your summons if you could procure one?"

"Nearest town, sir—anywhere."

The lady laughed heartily, and her little husband rubbed his hands and then patted her on the back.

"This lady is quite right, my dear Burne," said the professor. "I see that we shall be obliged to go armed."

"Armed, sir!—armed?"

"Yes. We shall for the greater part of our time be in places where the laws are of no avail, unless a body of troops are sent to enforce them."

"But then your firman will have furnished us with a Turkish soldier for our protection."

"But suppose the Turkish soldier prefers running away to fighting?" exclaimed the lady, "what then?"

"What then, ma'am?—what then?" cried the lawyer. "I flatter myself that I should be able to quell the people by letting them know that I was an English gentleman. Do you think that at my time of life I am going to turn butcher and carve folks with a sword, or drill holes through them with bullets?"

"Yes, sir, if it comes to a case of who is to be carved or drilled. There!—think it over. Come, Charley! let's have our walk."

Saying which the lady nodded and smiled to the two elders, and was going off in an assumed masculine way, when she caught sight of Lawrence lying back in an easy-chair, and her whole manner changed as she crossed to him and held out her hand with a sweet, tender, womanly look in her eyes.

"Good-bye for the present!" she said. "You must make haste and grow strong, so as to help me up the mountains if we meet somewhere farther in."

CHAPTER VI.

MR. BURNE TRIES A GUN.

"NOW that's just what I hate in women," said the old lawyer, viciously scattering snuff all over the place. "They put you in an ill temper, and rouse you up to think all sorts of bitter things, and then just as you feel ready to say them, they behave like that and disarm you. After the way in which she spoke to Lawrence there I can't abuse her."

"No, don't, please, Mr. Burne," said Lawrence warmly, and with his cheeks flushing, "I am sure she is very nice when you come to know her."

"Can't be," cried the lawyer. "A woman who advocates fire and sword. Bah!"

"But as a protection against fire and sword," said the professor laughing.

"Tchah, sir! stuff!" cried the other. "Look here; I can be pretty fierce when I like, and with you so big and strong, and with such a way with you as you have—Bah! nonsense, sir, we shall want no arms."

"Well, I propose that we now consult the landlord."

"Oh, just as you like, sir; but if he advocates such a proceeding, I'm not going to stalk through Turkey carrying fire-irons in my belt and over my shoulder, like a sham footpad in a country show."

The landlord was summoned—a frank-looking Eng-

lishman, who listened to all the professor said in silence
and then replied:

"Mr. Thompson the consul is quite right, sir. We
are not in England here, and though this is the nineteenth century the state of the country is terribly lawless. You know the old saying about when at Rome."

"Do as the Romans do, eh?"

"Exactly, sir. Every second man you meet here
even in the town goes armed, even if his weapons are
not seen, while in the country—quite in the interior,
it is the custom to wear weapons."

"Then I shall not go," said Mr. Burne decisively.

"If you ask my advice, gentlemen, I should say,
carry each of you a good revolver, a knife or dagger,
a sword, and a double-barrelled gun."

"Sword, dagger, and gun!" cried the professor.
"Surely a revolver would be sufficient."

"Why not push a nice large brass cannon before us
in a wheel-barrow?" said Mr. Burne sarcastically, and
then leaning back in his chair to chuckle, as if he had
said something very comical, and which he emphasized
by winking and nodding at Lawrence, who was too
much interested in the discussion upon weapons to
heed him.

"A revolver is not sufficient, for more than one reason,
gentlemen," said the landlord. "It is a deadly weapon
in skilful hands; but you will meet scores of people
who do not understand its qualities, but who would
comprehend a sword or a gun. You do not want to
have to use these weapons."

"Use them, sir? Of course not," roared the lawyer.

"Of course not, sir," said the landlord. "If you go

(348) D

armed merely with revolvers you may have to use them; but if you wear, in addition, a showy-looking sword and knife, and carry each of you a gun, you will be so formidable in appearance that the people in the different mountain villages will treat you with the greatest of respect, and you may make your journey in safety."

"This is very reasonable," said the professor.

"I assure you, sir, that in a country such as this is now such precautions are as necessary as taking a bottle of quinine. And beside, you may require your guns for game."

"The country is very fine, of course?"

"Magnificent, sir," replied the landlord; "but it is in ruins. The neglect and apathy of the government are such that the people are like the land—full of weeds. Why, you will hardly find a road fit to traverse, and through the neglect of the authorities, what used to be smiling plains are turned to fever-haunted marshes spreading pestilence around."

"You will have to give way, Mr. Burne," said the professor smiling, "and dress like a bandit chief."

"Never, sir," cried the lawyer. "You two may, but I am going through Asia Minor with a snuff-box and a walking-stick. Those will be enough for me."

"Where can we get arms?" said the professor smiling.

"At Politanie's, sir, about fifty yards from here. You will find him a very straightforward tradesman. Of course his prices are higher than you would pay in London; but he will not supply you with anything that is untrustworthy. Perhaps you may as well say that you are friends of our consul, and that I advised you."

"It is absurd!" exclaimed Mr. Burne, as soon as they were alone. "What do you say, Lawrence, my boy? You don't believe in weapons of war, I'm sure."

"No," replied Lawrence quietly.

"There, professor."

"But," continued Lawrence, "I believe in being safe. I feel sure that the people will respect us all the more for being armed."

"And would you use a sword, sir?" cried the lawyer fiercely.

Lawrence drew his sleeve back from his thin arm, gazed at it mournfully, and then looked up in a wistful half-laughing way at his two friends.

"I don't think I could even pull it out of the sheath," he said sadly.

"Come, Burne, you will have to yield to circumstances."

"Not I, sir, not I," said Mr. Burne emphatically. "I have been too much mixed up with the law all my life, and know its beauties too well, ever to break it."

"But you will come with us to the gunsmith's?"

"Oh, yes, I'll come and see you fool away your money, only I'm not going to have you carry loaded guns near me. If they are to be for show let them be for show. There, I'm ready."

"You will lie down for an hour, Lawrence, eh?" said the professor; "it is very hot." But the lad looked so dismayed that his friend smiled and said, "Come along, then."

A few minutes later they were in a store, whose

owner seemed to sell everything, from tinned meat to telescopes; and, upon hearing their wants, the shrewd, clever-looking Greek soon placed a case of revolvers before them of English and American make, exhibiting the differences of construction with clever fingers, with the result that the professor selected a Colt, and Lawrence a Tranter of a lighter make.

"He's a keen one," said Mr. Burne. "What a price he is asking for these goods!"

"But they seem genuine," said the professor; for the Greek had gone to the back of his store to make some inquiry about ammunition.

"Genuine fleecing," grumbled Mr. Burne; and just then the dealer returned.

"You select those two, then, gentlemen," he said in excellent English. "But if you will allow me, sir," he continued to Lawrence, "this is a more expensive and more highly finished pistol than the other, and it is lighter in the hand; but if I were you, as my arm would grow stronger, I should have one exactly like my friend's."

"Why?" said Lawrence; "I like this one."

"It is a good choice, sir, but it requires different cartridges to your friend's, and as you are going right away, would it not be better to have to depend on one size only? I have both, but I offer the suggestion."

"Yes, that's quite right," said the old lawyer sharply; "quite right. I should have both the same; and, do you know, I think perhaps I might as well have one, in case either of you should lose yours."

Mr. Preston felt ready to smile, but the speaker was

looking full at him, as if in expectation thereof, and he remained perfectly serious.

The pistols having been purchased, with a good supply of ammunition, guns were brought out, and the professor invested in a couple of good useful double-barrelled fowling-pieces for himself and Lawrence; Mr. Burne watching intently the whole transaction, and ending by asking the dealer to show him one.

"You see," he explained, "I should look odd to the people if I were not carrying the same weapons as you two, and besides I have often thought that I should like to go shooting. I don't see why I shouldn't; do you, Lawrence?"

"No, sir, certainly not," was the reply; and Mr. Burne went on examining the gun before him, pulling the lever, throwing open the breech, and peeping through the barrels as if they formed a double telescope.

"Oh! that's the way, is it?" he said. "But suppose, when the thing goes off, the shots should come out at this end instead of the other?"

"But you don't fire it off when it's open like that, Mr. Burne," cried Lawrence.

"My dear boy, of course not. Do you suppose I don't understand? You put in the cartridges like this. No, they won't go in that way. You put them in like that, and then you pull the trigger."

"No, no, no," cried Lawrence excitedly. "You shut the breech first."

"My dear boy—oh! I see. Yes, of course. Oh! that's what you meant. Of course, of course. I should have seen that directly. Now, then, it's all right. Loaded?"

"Sir! sir! sir!" cried the dealer, but he was too late, for
the old lawyer had put the gun to his shoulder, pointing the barrel towards the door, and pulled both triggers.

The result was a deafening explosion, two puffs of
smoke half filling the place, and the old gentleman was
seated upon the floor.

"Good gracious, Burne!" cried the professor, rushing
to him, "are you much hurt?"

Lawrence caught at the chair beside him, turning
ashy pale, and gazing down at the prostrate man,
while quite a little crowd of people filled the shop.

"Hurt?" cried Mr. Burne fiercely—"hurt? Hang
it, sir, do you think a man at my time of life can be
bumped down upon the floor like that without being
hurt?"

"But are you wounded—injured?"

"Don't I tell you, yes," cried Mr. Burne, getting up
with great difficulty. "I'm jarred all up the spinal
column."

"But not wounded?"

"Yes, I am, sir—in my self-respect. Here, help me
up. Oh, dear! Oh, lor'! Gently! Oh, my back! Oh,
dear! No; I can't sit down. That's better. Ah!"

"Would you like a doctor fetched?"

"Doctor? Hang your doctor, sir. Do you think I've
came out here to be poisoned by a foreign doctor. Oh,
bless my soul! Oh, dear me! Confound the gun! It's
a miserable cheap piece of rubbish. Went off in my
hands. Anyone shot?"

"No, sir," said the dealer quietly; "fortunately you
held the muzzle well up, and the charges went out of
the upper part of the door."

"Oh! you're there, are you?" cried Mr. Burne furiously, as he lay back in a cane chair, whose cushion seemed to be comfortable. "How dared you put such a miserable wretched piece of rubbish as that in my hands!"

The dealer made a deprecatory gesture.

"Here, clear away all these people. Be off with you. What are you staring at? Did you never see an English gentleman meet with an accident before? Oh, dear me! Oh, my conscience! Bless my heart, I shall never get over this.'

The dealer went about from one to the other of the passers-by who had crowded in, and the grave gentlemanly Turks bowed and left in the most courteous manner, while the others, a very motley assembly, showed some disposition to stay, but were eventually persuaded to go outside, and the door was closed.

"To think of me, a grave quiet solicitor, being reduced to such a position as this. I'm crippled for life. I know I am. Serves me right for coming. Here, give me a little brandy or a glass of wine."

The latter was brought directly, and the old lawyer drank it, with the result that it seemed to make him more angry.

"Here, you, sir!" he cried to the dealer, who was most attentive; "what have you to say for yourself? It's a wonder that I did not shoot one of my friends here. That gun ought to be destroyed."

"My dear Burne," said the professor, who had taken the fowling-piece and tried the locks, cocking and re-cocking them over and over again; "the piece seems to me to be in very perfect order."

"Bah! stuff! What do you know about guns?"

"Certainly I have not used one much lately, and many improvements have been made since I used to go shooting; but still I do know how to handle a gun."

"Then, sir," cried the little lawyer in a towering fury, "perhaps you will be good enough to tell me how it was that this confounded piece of mechanism went off in my hands?"

"Simply," said the professor smiling, "because you drew both the triggers at once."

"It is false, sir. I just rested my fingers upon them as you are doing now."

"And the piece went off!" said the professor drily, but smiling the while "It is a way that all guns and pistols have."

The dealer smiled his thanks, and Mr. Burne started up in the chair, but threw himself back again.

"Oh, dear! oh, my gracious me!" he groaned; "and you two grinning at me and rejoicing over my sufferings."

"My dear sir, indeed I am very sorry," said the dealer.

"Yes, I know you are," said Mr. Burne furiously, "because you think, and rightly, that I will not buy your precious gun. Bless my heart, how it does hurt! I feel as if I should never be able to sit up again. I know my vertebræ are all loose like a string of beads."

"Will you allow us to assist you into my private room, sir?" said the dealer.

"No, I won't," snapped the sufferer.

"But there is a couch there, and I will send for the resident English doctor."

"If you dare do anything of the kind, confound you, sir, I'll throw something at you. Can't you see that there is nothing the matter with me, only I'm in pain."

"But he might relieve you, Burne," said the professor kindly.

"I tell you I don't want to be relieved, sir," cried the little lawyer. "And don't stand staring at me like that, boy; I'm not killed."

"I am afraid that you are a great deal hurt," said Lawrence, going to his side and taking his hand.

"Oh, dear! oh, dear!" groaned the sufferer. "Well, I'm not, boy, not a bit. There."

"Let me send for a doctor, sir," said the dealer.

"I tell you I will not, man. Do you take me for a Greek or a Turk, or a heretic? Can't you see that I am an Englishman, sir, one who is never beaten, and never gives up? There, go on selling your guns."

"Oh, nonsense!" said the professor; "we cannot think of such things with you in that state."

"State? What state, sir? Here you, Mr. What's-your-name, I beg your pardon. I ought to have known better. Not used to guns. Pens are more in my way. Confoundedly stupid thing to do. But I've learned more about a gun now than I should have learned in six months. I beg your pardon, sir."

"Pray, say no more, sir," replied the dealer; "it is not needed."

"Yes, it is, sir," cried the lawyer fiercely. "Didn't I tell you I was an English gentleman. An English gentleman always apologizes when he is in the wrong. I apologize. I am very sorry for what I said."

The dealer smiled and bowed, and looked pleased as he handed the sufferer another glass of wine, which was taken and sipped at intervals between a few mild *ohs!* and *ssfths!*

"Not a bad wine this.   What is it?"

"One of the Greek wines, sir."

"Humph! not bad; but not like our port.   Now, you people, go on with your business, and don't stare at me as if I were a sick man.   Here, Mr. What's-your-name, put that gun in a case, and send it round to the hotel.   I've taken a fancy to it."

"Send—this gun, sir?"

"Yes.   Didn't I speak plainly?   Didn't the professor, my friend here, say it was a good gun?"

"Yes, sir, yes: it is an excellent piece of the best English make."

"Well, I want a gun, and I suppose any piece would go off as that did if somebody handled it as stupidly as I did."

"Yes, sir, of course."

"Then send it on, and the pistol too.   Ah, that's better—I'm easier; but I say, Preston, I shall have to be carried back."

"I'm very glad you are easier, but really if I were you I would see a doctor."

"I've no objection to seeing a doctor, my dear sir, but I'm not going to have him do anything to me."

"Then you really wish us to go on with our purchases?"

"Why, of course, man, of course.   What did we come for?   Go on, man, go on.   Here, mister, show me one or two of these long carving knives."

"Carving knives?" said the dealer. "I do not keep them."

"Yes, you do: these," said Mr. Burne, pointing to a case in which were several Eastern sabres.

"Oh, the swords!" said the dealer smiling. "Of course."

"You are not going to buy one of these, are you, Mr. Burne?" said Lawrence eagerly.

"To be sure I am," was the reply. "Why shouldn't I play at soldiers if I like. There, what do you say to that?" he continued, drawing a light, keen-looking blade from its curved sheath. "Try it. Mind it don't go off—I mean, don't go slashing it round and cutting off the professor's legs or my head. Can you lift it?"

"Oh, yes," cried Lawrence, poising the keen weapon in his hand before examining its handsome silver inlaid hilt.

"Think that would do for me? Oh, dear me, what a twinge!"

"Yes, sir, admirably," replied Lawrence.

"Then I don't," was the gruff retort. "Seems to me that it would just suit you. There, buckle on the belt."

Lawrence did as he was told, but the belt was too large and had to be reduced.

"Hah! that's better," said Mr Burne. "There, that's a very handsome sword, Lawrence, and it will do to make you look fierce when we are in the country, and to hang up in your room at home to keep in memory of our journey. Will you accept it, my boy, as a present?"

"Oh, thank you," cried the lad excitedly.

"Took a fancy to it as soon as you saw it, you young

dog. I saw you!" cried the old lawyer chuckling. "There, now for a dagger or knife to go with it."

The dealer produced one in an ornamental sheath directly, and explained that it was for use as a weapon, for hunting, or to divide food when on a journey.

"That will do, then, nicely. There, my boy, these are my presents. Now, Preston, I suppose we must each have one of these long choppers?"

"Yes, I think so," replied the professor. "They will make us look more formidable."

"Very well, then: choose one for me too, but I warn you, I shall fasten mine down in the sheath with gum. I'm not going to take mine out, for fear of cutting off somebody's legs or wings, or perhaps my own."

"You feel better now?" said the professor.

"Hold your tongue, sir — do! No: I don't feel better. I had forgotten my pain, but now you've made me think about it again. There!—choose two swords and knives and let's get back."

Two plain useful sabres were selected, and the dealer received his orders to send the weapons to the hotel, after which the injured man was helped into a standing position, but not without the utterance of several groans. Then he was walked up and down the shop several times, ending by declaring himself much better.

"There, Lawrence!" he cried, "that's the advantage of being an Englishman. Now, if I had been a Dutchman or a Frenchman I should have had myself carried back, sent for a couple of doctors, and been very bad for a month or two; but you see I'm better already, and I'm not going to give up to please the Grand Panjandrum himself. Dear me! bless my heart! Panjandrum! Pan — pan — pan — jan — jan — jan — drum! Where did I hear that word?"

"In a sort of nursery ditty, sir," said Lawrence laughing.

"To be sure I did," cried the old man, "and I had forgotten it; but I say, don't laugh like that, boy."

"Why not, sir?"

"Because it will make us believe that you have been shamming all this time, and that you're really quite well, thank you, sir!—eh?"

"I—I think I am better," said Lawrence quickly. "I don't know why, but I have not been thinking about being ill these last few days, everything is so bright and sunshiny here, you see."

"Yes, I see," said the old lawyer, giving the professor a peculiar look; and they went back to the hotel.

CHAPTER VII.

THE GREEK SKIPPER.

O, I can't do it," said Mr. Burne after several brave efforts; "I really am a good deal jarred, and it is quite impossible. I am quite right as long as I keep still, but in such pain if I move that I can hardly bear it."

"Then we will put off the journey for a week," said the professor decisively.

"And disappoint the lad?" said Mr. Burne. "No; you two must go."

"How can you talk like that?" exclaimed Lawrence sharply, "when you have come on purpose to help me get strong again? Mr. Preston, we shall stay here—shall we not?"

"Of course," replied the professor. "The enjoyment of our trip depends upon our being staunch to one another."

Mr. Burne declared that it was absurd, and ridiculous, and nonsensical, and raked out a few other adjectives to give force to his sentiments, speaking in the most sour way possible; but it was very evident that he was highly pleased, and the steamer sailed without them.

The next day Mr. Burne was so stiff that he could not walk about; but he refused to see a doctor, and a week passed before he could move without pain. Then one morning he declared that he was mending fast, and insisted upon inquiries being made respecting the sailing of the next steamer that would stop at one or other of the little towns on the south coast; but there was nothing bound in that direction, nor likely to be for another fortnight.

"And all my fault!" cried Mr. Burne angrily. "Tut-tut-tut! Here, ring for the landlord."

The landlord came and was questioned.

No, there was no possibility of a passage being made for quite a fortnight, unless the visitors would go in a small sailing boat belonging to one or the other of the trading crews.

The professor glanced at Lawrence, thought of the probable discomfort, and shook his head.

"The very thing!" exclaimed Mr. Burne sharply.

"We can make trips in steamers at any time; but a trip in a Greek felucca, with real Greek sailors, is what I have longed for all my life. Eh, Lawrence, what do you say?"

"I think with you, sir, that it would be delightful—that is, if you are well enough to go."

"Well enough to go! of course I am. I'm longing to be off. Only a bit stiff. Look here, landlord, see what you can do for us. One moment, though; these Greeks—they will not rob us and throw us overboard—eh?"

"No fear, sir. I'll see that you go by a boat manned by honest fellows who come regularly to the port. Leave it to me."

The landlord departed and the question was discussed. The professor was ready enough to go in the manner proposed so long as Lawrence felt equal to the task, and this he declared he was; and certainly, imperceptibly as it had come about, there was an improvement in his appearance that was most hopeful.

The principal part of their luggage had gone on by steamer, and would be lying waiting for them at Ansina, a little port on the south coast which had been considered a suitable starting-point; and they had been suffering some inconvenience, buying just such few things as would do to make shift with till they overtook their portmanteaux.

Oddly enough, Mr. Burne expressed the most concern about their new purchases, the weapons and ammunition, which had been sent on to the steamer by the landlord as soon as they arrived from the store.

"Such things must be so tempting to the people who see them," said the old lawyer.

"But they were all carefully packed in cases," said the professor. "They would not know what was inside."

"Nonsense, my dear sir. We English folk would not have known, but a Greek or a Turk would. These people smell powder just like crows in a corn-field. I'm afraid that if we don't make haste we shall find our things gone, and I wouldn't lose that gun for any money."

The landlord came back in about a couple of hours to say that he had had no success, but that it would become known that he had been inquiring, and an application might be made.

This turned out correct, for as the travellers were seated that evening over their dessert, enjoying by an open window the deliciously soft breeze, as Lawrence partook of the abundant grapes, and the professor puffed at a water-pipe—an example followed by Mr. Burne, who diligently tried to like it, but always gave up in favour of a cigar at the end of a quarter of an hour—the waiter brought their coffee and announced that the master of a small vessel desired to see their excellencies.

The man was shown in, and proved to be a picturesque-looking fellow in a scarlet cap, which he snatched from his curly black hair and advanced into the room, saying some words in modern Greek whose import the professor made out; but his attempts to reply were too much for the skipper, who grew excited, shook his head, and finally rushed out of the room, to the great amusement of Mr. Burne, who knocked the ash off the cigar he had recently lit.

"That's what I always say," he cried. "Book language is as different as can be from spoken language. I learned French for long enough when I was a boy, but I never could make a Frenchman understand what I meant."

"Let's ring and inquire," said the professor, to hide a smile. "I hope we have not driven the fellow away."

"Hope you have, you mean," said Mr. Burne.

The professor rose to reach the bell, but just then the landlord entered with the Greek sailor, who smiled and showed his white teeth.

With the landlord as interpreter the matter became easy. The man was going to sail in three days, that was as soon as the little vessel, in which he had brought a cargo of oranges and other fruit from Beyrout, had discharged her load and was ready to return. He was going to Larnaca on his return voyage, but for a consideration he was ready to take the English excellencies to any port they liked on the south coast — Ansina if they wished — and he would make them as comfortable as the boat would allow; but they must bring their own food and wine.

The bargain was soon struck, the Greek asking a sum which the landlord named to the professor—so many Turkish pounds.

"But is not that a heavy price for the accommodation we shall receive?"

"Very," said the landlord smiling. "I was going to suggest that you should offer him one-third of the amount."

"Then we shall offend him and drive him away," said Mr Burne.

"Oh, dear me! no, gentlemen. He does not expect to get what he asks, and the sum I name would be very fair payment. You leave the settlement in my hands."

The professor acquiesced, and the landlord turned to the Greek sailor to offer him just one-third of the sum he had asked.

"I thought as much," said the old lawyer. "The landlord thinks we're in England, and that it was a bill of costs that he had to tax. Look at the Greek, Lawrence!"

The latter needed no telling, for he was already watching the sailor, who was protesting furiously. One moment his hands were raised, the next they were clenched downwards as if about to strike the floor. Again they were lifted menacingly, and there seemed danger, for one rested upon a knife in his belt, but only for it to be beaten furiously in the other. Quick angry words, delivered with the greatest volubility, followed; and then, turning and looking round in the most scornful manner, the man seemed to fire a volley of words at the whole party and rushed from the room.

"I'm sorry for this," said the professor, "for we would have paid heavily sooner than wait longer."

"Humph! Yes," exclaimed Mr. Burne. "Why not call the man back and offer him two-thirds of his price?"

"Because, sir," replied the landlord, "it would have been giving him twice as much as would pay him well. Don't you see, sir, that he is going back empty, and every piastre you pay him is great profit. Besides, I presume that you will take far more provisions than will suffice for your own use."

. " Naturally," replied the professor.

" And this man and his little crew will reap the benefit?"

" But you have driven him away."

" Oh dear, no, sir!" replied the landlord smiling. " He will be back to-night, or at the latest to-morrow morning, to seal the bargain."

" Do you think so?" cried Lawrence, who looked terribly disappointed at this new delay.

" I am sure," said the landlord laughing. " Here he is."

For there was a quick step on the stair, the door was opened, and the swarthy face of the Greek was thrust in, the red cap snatched off, and, showing his white teeth in a broad smile, he came forward, nodding pleasantly to all in turn.

A few words passed, the bargain was made, and the tall lithe fellow strode out in high glee, it being understood that he was to well clean out the little cabin, and remove baskets and lumber forward so as to make the boat as comfortable as he could for his passengers; that he was to put in at any port they liked, or stop at any island they wished to see; and, moreover, he swore to defend them with his men against enemies of every kind, and to land them safely at Ansina, or suffer death in default.

This last was his own volunteered penalty, after which he darted back to say that their excellencies might bring a little tobacco for him and his men, if they liked, and that, in return, they might be sure of finding a plentiful supply of oranges, grapes, and melons for their use.

"Come, landlord," said Mr. Burne, "I think you have done wonders for us."

"I have only kept you from being cheated, gentlemen," was the reply. "These men generally ask three or four times as much as they mean to take."

"And do the landlords?" said the professor drily.

"I hope not, sir," was the reply. "But now, gentlemen, if you will allow me, I should like to offer you a bit of advice."

"Pray, give it," said the professor gravely.

"I will, sir. It is this. You are going into a very wild country, where in places you will not be able to help yourselves in spite of your firman. That will be sufficient to get you everything where the law is held in anything like respect, but you will find yourselves in places where the rude, ignorant peasants will look upon you as Christian dogs, and will see you starve or die of exposure before they will give or even sell you food for yourselves or horses."

"Mighty pleasant set of barbarians to go amongst, I must say!" cried Mr. Burne.

"I am telling you the simple truth, gentlemen. You will find no hotels or inns, only the resting-places—the khans—and often enough you will be away from them."

"He is quite right," said the professor calmly. "I was aware that we should sometimes have to encounter these troubles."

"Humph! 'Pon my word!" grumbled Mr. Burne. "Look here, Lawrence, let's go back."

" What for?" cried the lad flushing. "Oh, no! we must go on."

The professor glanced at him quickly, and smiled in his calm grave way before turning to the landlord.

"You have not given us your advice," he said.

"It is very simple, gentlemen, and it is this: Take with you a man who knows the country well, who can act as guide, and from his frequent travels there can speak two or three languages—a faithful trusty fellow who will watch over you, guard you from extortion, and be ready to fight, if needs be, or force the people he comes among to give you or sell you what you need."

"Oh! but are they such savages as this—so near to the more civilized places of the East?"

"Quite, sir," replied the landlord.

"And where is this pearl among men to be found?" said the professor with a slight sneer. "Do you know such a one?"

"Yes, sir; he only returned from a journey yesterday. I happened to see him this morning, and thought directly of you."

"Would he go with us?" said the old lawyer quickly.

"I cannot say for certain," was the reply; "but if you will give me leave I will see him and sound him upon the subject."

"Humph!" from the old lawyer.

"He has just been paid, and would no doubt like to stay and rest here a little while, but I daresay I could prevail upon him to go with you if he saw you first."

"Then he is to be the master, not we?"

" Well, gentlemen, I don't say that," said the landlord

smiling; "but people out here are very different to what they are at home. I have learned by bitter experience how independent they can be, and how strong their natural dislike is to Christians."

"This man is not a Christian, then?"

"Oh, no, sir! a Muslim, a thorough-going Turk."

"He will not carry his religious feelings to the pitch of pushing us over some precipice in the mountains, eh? and then come home thinking he has done a good work, eh, Mr. Landlord?" said the old lawyer.

"Oh, no! I'll answer for his integrity, sir. If he engages to go with you, have no hesitation in trusting him with your baggage, your arms, your purses if you like. If he undertakes to be your guide, he will lose his life sooner than see you robbed of a single piastre."

"And what will he require?" said Mr. Burne shortly; "what pay?"

"Very moderate, gentlemen, and I promise you this, that if I can persuade him to go with you, the cost of paying him will be saved out of your expenses. I mean that you will spend less with him than you would without."

"And he knows something of the country?"

"A great deal, gentlemen. Shall I see if I can get him to go?"

"By all means," cried the two elders in a breath.

"If he consents I will bring him to you. I beg pardon, I am wrong. I must bring him to see you first before he will consent."

"Then, as I said before, he is to be the master, not we," said the professor.

"No, no, sir, you must not take it like that. The

man is independent, and need not undertake this journey without he likes. Is it surprising, then, that if he should come and see you, and not liking your appearance, or the prospect of being comfortable in your service, he should decline to go?"

"You are quite right," said Mr. Burne. "I would not."

CHAPTER VIII.

YUSSUF THE GUIDE.

A'T' breakfast-time the next morning the land-lord came and announced that Yussuf was in waiting. A few minutes later he ushered in a rather plain-looking, deeply-bronzed, middle-aged man, who, at the first glance, seemed to have nothing whatever to recommend him. As a nation his people are good - looking and dignified. Yussuf was rather ill-looking and decidedly undignified. He did not seem muscular, or active, or clever, or agreeable, or to have good eyes. He was not even well dressed. But upon further examination there was a hardened wiry look about the man, and a stern determined appearance in the lines of his countenance, while the eyes that did not seem to be good, so sunken were they beneath his brow, and so deeply shaded, were evidently keen and piercing. They seemed to flash as they met those of the old lawyer, to look defiant as they encountered the professor's searching gaze, and then to soften as they were turned upon Lawrence, as he lay back in his chair rather exhausted by the heat.

A few questions were asked on either side, the new-comer speaking very good English, and also grasping the professor's Arabic at once. In fact, it appeared evident that he was about to decline to accompany the party; but the words spoken sonorously by the professor seemed to make him hesitate, as if the fact of one of the party speaking the familiar tongue gratified him, but still he hesitated.

Just then, he hardly knew why, but attracted by the eyes of the Turk, which were fixed upon him gravely, and in a half-pitying manner, Lawrence rose and approached.

"I hope you will go with us," he said quickly.

Yussuf took his hand and held it, gazing in the lad's face earnestly, as a pleasant smile illumined his own.

"You are weak and ill," he said softly. "The wind that blows in the mountains will make you strong."

Then turning slowly to the others he saluted them gravely.

"Effendis," he said, "I am thy servant. Allah be with us in all our journeyings to and fro. I will go."

"I am glad!" cried Lawrence.

"And so am I," said the professor, hesitating for a moment, and then holding out his hand, which Yussuf took respectfully, held for a moment, and then turned to Mr. Burne.

"Oh, all right, shake hands," said the latter, "if it's the custom of the country; and now about terms."

"Leave me to settle that with Yussuf," said the landlord hastily, and he and the Muslim left the room.

THE ENGAGEMENT OF YUSSUF THE GUIDE.

"Seems queer to begin by being inspected, and then shaking hands with the servant we engage, eh, professor?" said Mr. Burne.

"The man is to be more than servant," replied Mr. Preston; "he is to be our guide and companion for months. He repelled me at first, but directly he spoke in that soft deep voice there seemed to me to be truth in every accent. He is a gentleman at heart, and I believe we have found a pearl. What do you say, Lawrence?"

"He made me like him directly he looked in my eyes, and I am very glad he is going."

"I repeat my words," said the professor.

"Well, I mustn't quarrel, I suppose. My back's too bad; so I throw in my lot with you, and say I am glad, and good luck to us."

"Amen," said the professor gravely; "but I like our guide's way of wishing success the better of the two."

CHAPTER IX.

YUSSUF IS SUSPICIOUS.

LAWRENCE watched anxiously for the arrival of the new guide Yussuf on the day appointed for sailing. There had been one more disappointment, the Greek having declared that he must have another day before he would be ready, but there was no further delay.

Yussuf came to say that he had examined the boat,

that it was good, seaworthy, and well manned by a stout little crew of sailors, but that he was very much dissatisfied with the accommodation prepared for the gentlemen.

He had not been told to report upon this matter, and his evident quiet eagerness to serve his employers well was satisfactory.

" We expect to rough it," said the professor. " It will not be for long."

Yussuf shrugged his shoulders, and said as he looked hard at Lawrence:

" It may be long, effendi. The winds perhaps light, and there are storms."

" I am afraid we must risk these troubles; and be- sides, it is a coasting trip, and we should be able to run into some port."

Yussuf bowed.

" I thought it my duty to tell his excellency of the state of the boat," he said; and then, in an earnest busy way, he asked about the baggage to go on board, and provisions, promising to bring up a couple of the Greek sailors to carry down what was necessary.

In the course of the afternoon this was done, the consul visited and parted from in the most friendly manner, Lawrence's eyes brightening as the official rested his hand upon his shoulder, and declared in all sincerity that he could see an improvement in him already.

The landlord endorsed this remark too on parting, and he as well as the consul assured the little party that, if anything could be done to help them, a message would receive the most earnest attention.

"You think we shall get into trouble, then?" Lawrence ventured to say, but shrank back directly he had spoken, with his cheeks flushed and heart beating, for his long illness had made him effeminate.

"I think it possible," said the landlord smiling; "but I sincerely hope you will not. In fact, with a man like Yussuf your risks are greatly reduced. Good-bye, gentlemen, and I shall look forward to seeing you again on your way back."

"Bravo, Lawrence!" cried the professor, clapping him on the shoulder. "I had been thinking the same thing; now I am sure of it."

"I don't understand you," said the lad wiping his face, for the perspiration was standing in a fine dew all over his brow.

"Why, both Mr. Thompson and the landlord here said that you were better, and you have just shown me that you are."

"How, Mr. Preston?" said the lad bashfully.

"By the way in which you just now spoke out, my boy," said Mr. Burne, joining in. "Why, you couldn't have spoken like that before we started. You are not much better now; but when we settled to come on this trip you were as weak and bashful as a delicate girl. Preston, we shall make a man of him after all."

They were walking towards the landing-place nearest to where the Greek's boat lay, and further conversation was stayed by Yussuf coming to them.

'The boatman will not believe, excellencies," he said, "that there is no more luggage. Have I got all?"

"Yes; all our luggage went on by the steamer to Ansina."

Yussuf bowed and went back to the landing-place, where a small boat manned by the Greek and one of his men was in waiting, and in the travellers' presence Yussuf explained about their belongings.

The Greek listened with rather a moody expression, but said no more; and in a very short time the little party were pulled to the side of a long light craft, about the burden of a large west country fishing lugger, but longer, more graceful in shape, and with the fore-part pretty well cumbered with baskets, which exhaled the familiar ether-like odour of oranges.

The accommodation was very spare, but, as the weather was deliciously fine, there was little hardship in roughing it in the open—provision being made for the invalid to stay in shelter as much as he liked.

They began to find the value of their guide at once, for he eagerly set to work to find them seats by improvising places in the stern; showing how he had arranged the provisions and fresh water, and offering Lawrence some ripe grapes as he made him comfortable where he would be out of the way of the men hoisting sail, and getting clear of the many boats lying at hand. First one and then the other long tapering sail was hoisted, each looking like the wing of a swallow continued to a point, as it stretched out to the tip of the curved and tapering spar; and as these filled the light vessel careened over, and began to glide swiftly through the bright blue sea.

After lending some help the Greek skipper went behind his passengers to the helm, his crew of three swarthy-looking fellows, each with his knife in his belt, threw themselves down amongst the baskets forward, and as the passengers stood or sat watching the glorious panorama of town, coast, and shipping they were passing, Yussuf calmly shook his loose garment about him, squatted down beside the low bulwark, and lighting a water-pipe began to smoke with his eyes half closed, and as if there was nothing more to trouble about in life.

"'Pon my word!" said the old lawyer. "What a place this boat seems to be for practising the art of doing nothing comfortably!"

"Yes," said the professor, taking in the scene on board at a glance. "It is typical of the East. You must get westward to see men toiling constantly like ants. The word business does not belong to these lands."

"You are right," said Mr. Burne.

"Well, it is the custom of the country," continued the professor, "and while we have no hard travel to do, let us follow these people's example, and watch and think."

"There is no room to do anything else," said Mr. Burne grumpily.

"How delicious!" said Lawrence as if to himself.

"What, those grapes!" said the professor smiling.

"I beg your pardon!" exclaimed Lawrence, starting and flushing again like a girl. "No; I meant sitting back here, and feeling this beautiful soft breeze as we glide through the blue sea."

"You like it then?" said Mr. Burne smiling.

"Oh, yes! I don't know when I felt so well and happy. It is delightful."

"That's right," cried Mr. Burne. "Come, now; we must throw the invalid overboard."

Lawrence laughed.

"I mean the disease," said Mr. Burne. "No more talking about being ill."

"No," said Lawrence quietly, and speaking as if he felt every word he uttered to be true; "I feel now as if I were growing better every hour."

"And so you are," cried the professor. "Come, don't think about yourself, but set to work and take photographs."

"Nonsense!" cried Mr. Burne; "let the boy be, now he is comfortable. Photographs indeed! Where's your tackle?"

"I mean mental photographs," said the professor laughing.

"Then, why didn't you say so, man? Good gracious me, if we lawyers were to write down one thing when we mean another, a pretty state of affairs we should have. The world would be all lawsuits. Humph! Who'd think that Smyrna was such a dirty, shabby place, to look at it from here?"

"A lovely scene certainly!" said the professor. "Look, Lawrence, how well the mountain stands out above the town."

"Humph, yes; it's very pretty," said the lawyer; "but give me Gray's Inn with its plane-trees, or snug little Thavies' Inn. This place is a sham."

"But it is very beautiful seen from here, Mr. Burne," said Lawrence, who was feasting on the glorious sunlit prospect.

"Paint and varnish, sir, over rotten wood," snorted Mr. Burne. "Look at the drainage; look at the plagues and fevers and choleras they get here."

"Yes," said the professor, "at times."

"Bah! very pretty, of course, but nothing like London."

"With its smoke," said the professor.

"Fine healthy thing, sir," cried the old gentleman. "Magnificent city, London!"

"And its darkness and fogs," said Lawrence.

"Well, who minds a bit of fog, so long as he is well?" cried Mr. Burne. "Look here, young man; don't you find fault with your own land. Stick up for it through thick and thin."

"For all of it that is good, my lad," said the professor merrily, "but don't uphold the bad."

"Bad, sir! There's precious little that's bad in London. If you want to go a few hundred miles there, you can go at any time and get good accommodation. Not be forced to ride in a market-boat with hard seats. Bless me, they are making my back bad again."

"Oh, but, Mr. Burne, look, look, the place here is lovely!"

"Oh, yes, lovely enough, but, as the fellow said, it isn't fit to live in long; it's dangerous to be safe."

"What do you mean?"

"Earthquakes, sir. If you take a house in London, you know where you are. If you take one here, as the fellow said, where are you? To-day all right, to-morrow shaken down by an earthquake shock, or swallowed up."

"There are risks everywhere," said the professor, who seemed to be gradually throwing off his dreamy manner, and growing brighter and more active, just as if he had been suffering from a disease of the mind as Lawrence had of the body.

"Risks? Humph! yes, some; but by the time we've finished our trip, you'll all be ready to say, There's no place like home."

"Granted," said the professor.

"Why, you're not tired of the journey already, Mr. Burne?"

"Tired? No, my boy," cried the old man smiling. "I'm in a bad temper to-day, that's all. This seat is terribly hard and—oh, I know what's the matter. I'm horribly hungry."

He turned his head to see that Yussuf had finished and put away his pipe, and was busy over one of the baskets of provisions, from which he produced a cloth and knives and forks, with a bottle of wine and several other necessaries, which his forethought had suggested; and in a short time the travellers were enjoying a rough but most palatable *al fresco* meal in the delicious evening, with the distant land glowing with light of a glorious orange, and the deep blue sea dappled with orange and gold.

"We have plenty of provisions, I suppose," said the professor.

"Yes, effendi, plenty," said Yussuf, who had been taking his portion aside.

"Then pass what is left here to the skipper and his men."

Yussuf bowed gravely, and the men, who had been making an evening meal of blackish bread and melons, were soon chattering away forward, eating the remains of the meal and drinking a bottle of the Greek wine Lawrence took them.

The tiller had been lashed so as to set the Greek

skipper at liberty, and the travellers were alone, while, wearied by his extra exertion, Lawrence lay back, apparently fast asleep, when Yussuf approached the professor and his companion, with his water-pipe which he was filling with tobacco, and about which and with a light, he busied himself in the most matter-of-fact manner.

But Yussuf was thinking of something else beside smoke, for he startled the professor and made Mr. Burne jump and drop his cigar, as he said in a low voice:

"Your excellencies are well armed, of course?"

"Armed?" exclaimed the professor.

Yussuf did not speak, but stooped to pick up the fallen cigar, which he handed to its owner.

"Be calm, excellency," he said smiling, "and tell me."

The professor looked at him suspiciously; but there was that in the man's countenance that disarmed him, and he said quietly:

"We certainly have plenty of arms."

"That is good," said Yussuf, with a flash of the eye.

"But our weapons are packed up with our luggage, and went on by the steamer."

"That is bad," said Yussuf quietly.

"We never thought they would be necessary till we got ashore."

"Look here, my man," said Mr. Burne; "speak out. Are you suspicious of these people?"

"My life has taught me to be suspicious, effendi," said Yussuf, lighting his pipe, "particularly of the low-class Greeks. They are not honest."

"But surely," began the professor.

(348)   F

"Be perfectly calm, effendi," said Yussuf, pointing shoreward, and waving his hand as if telling the name of some place. I have nothing certain against this Greek and his men; but we are out at sea and at their mercy."

"But something has happened to make you speak like this," said Mr. Burne with a searching look.

"A trifle, effendi," replied the Muslim; "but a little cloud like that yonder"—pointing seaward now beyond the Greek sailors, so that the travellers could see that they were watched by the skipper—"is sometimes the sign of a coming storm."

"Then what have you seen?" said Mr. Burne suspiciously.

"A trifle—almost nothing, effendi, only that the man there was out of temper when he found that all your baggage had gone."

"Humph!" ejaculated Mr. Burne.

"Then you think there is danger?" said the professor.

"I do not say that," said Yussuf, pointing shoreward again, "but your excellencies may as well learn your lessons at once. We are commencing our journey, and are now, as we generally shall be, at the mercy of men who obey the laws when they feel the rod over their backs, but who, when they cannot see the rod, laugh at them."

"What do you ask us to do, then?" said the professor quickly.

"Be always on guard, but never show it. Be prepared for danger. If there is none, so much the better. Life here is a little matter compared to what I am told it is among you Franks, and it becomes every man's duty to guard his life."

"But these Greek sailors?" said Mr. Burne sharply.

"I do not trust them," replied Yussuf calmly. "If we are the stronger they will be our slaves. If they feel that they are, our lives would not be safe if they had the chance to rob us. They believe your excellencies to be rich and to have much gold."

"Look here, Yussuf," said Mr. Burne uneasily, "our friend ashore gave you a capital character."

"I have eaten salt with your excellencies, and my life is yours," replied Yussuf.

"Then what would you do now?"

"Be perfectly calm, effendi, and treat these men as if you did not know fear."

"And we have no arms," said Mr. Burne uneasily.

"Can your excellency fight?" said Yussuf quietly.

"A law case—yes, with any man, but any other case of fighting—good gracious me, no. I have not fought since I had a black eye at school."

"But you can, effendi?" continued Yussuf, looking with admiration at the professor's broad chest and long muscular arms.

"I daresay I can, if I am driven to it," replied the professor gravely; and he involuntarily clenched a large, hard, bony hand.

"Yes," said Yussuf, with a grave smile of satisfaction. "Your excellency can fight, I see."

"But we are entirely without arms," repeated Mr. Burne excitedly.

"Not quite," replied Yussuf calmly. "Your excellency has a big stick; the effendi here has hands and

strength that would enable him to throw an enemy
into the sea, and I never go a journey without my
pistol and a knife."

"You have a pistol?" said Mr. Burne eagerly.

"Be quite calm, excellency," said Yussuf, laughing
as he smoked, and bowing down as if something droll
had been said. "Yes, I have a pistol of many barrels
given to me by a Frankish effendi when we returned
from a journey through the land of Abraham, and
then down to the stony city in the desert — Petra,
where the Arab sheiks are fierce and ready to rob all
who are not armed and strong."

"Where is it?" said the professor.

"Safe in my bosom, effendi, where my hand can
touch it ere you blink an eye. So you see that we
are not quite without arms. But listen," he continued;
"this may be all a fancy of mine."

"Then you will do nothing?" exclaimed Mr. Burne.

"Oh no, I do not say that, effendi. We must be
watchful. Two must sleep, and two must watch night
or day. The enemy must not come to the gate and
find it open ready for him to enter in."

"Those are the words of wisdom," said the professor
gravely, and Yussuf's eyes brightened and he bowed.

"This watchfulness," he said, "may keep the enemy
away if there be one. If there be none: well, we have
taught ourselves a lesson that will not be thrown
away."

"Why, Yussuf, I am beginning to think you are a
treasure!" exclaimed Mr. Burne.

Yussuf bowed, but he did not look pleased, for he
had not warmed towards the old lawyer in the slightest

degree. He had been met with distrust, and he was reserved towards him who showed his doubt so openly.

"I thought it was but just, effendis, to warn you, and I thought it better to say so now, while the young effendi is asleep, for fear he might be alarmed."

"I am not asleep," said Lawrence turning his head. "I have not been to sleep."

"Then you have heard all that was said," exclaimed the professor.

"Every word, Mr. Preston. I could not help hearing," said Lawrence, sitting up with his face flushed and eyes brightened. "I did not know till just now that I was not expected to hear."

"Humph, and do you feel alarmed?" said the old lawyer.

"I don't think I do, sir," replied the lad calmly. "Perhaps I should if—if there should be a fight."

"I do not think there will be," said the professor quietly. "Yussuf here has warned us, and forewarned is forearmed."

"Even if we have no pistols, eh?" said Mr. Burne laughing, but rather acidly. "Humph, here comes the skipper."

The Greek came aft smiling and unlashed the tiller, altering their course a little, so that as the evening breeze freshened they seemed literally to skim along the surface of the sea.

CHAPTER X.

A NIGHT OF HORRORS.

HE night came, with the stars seeming to blaze in the clear atmosphere. The skipper had given up the helm to one of his men, and joined the others forward to lie down among the baskets and sleep, as it seemed, while aft, at the professor's request, Mr. Burne and Lawrence lay down to sleep, leaving the others to watch.

The night grew darker, and the water beat and rippled beneath the bows, all else being wonderfully still as the boat glided on.

Yussuf lit his water-pipe, and the professor a cigar, to begin conversing in a low tone, but always watchful of the slightest movement of the men.

A couple of hours had glided away, and then, after being apparently fast asleep, the skipper rose and came aft to speak eagerly to Yussuf, who heard him out, and then turned to the professor.

"The captain says that there is no danger of wreck or storm; that he and his men will watch over you as if you were given over to their safe keeping, and all will be well."

"Tell the captain that I prefer to sit up and watch the sea and sky," replied the professor. "When I am tired I will lie down."

The skipper nodded and smiled, and went forward again, while, after some minutes' silence, the professor said softly:

"You are quite right to be doubtful, Yussuf, I mistrust that man."

"Yes," replied Yussuf in the same tone, "the Greek dog will bite the hand which fed him if he has a chance, but that chance, effendi, he must not have."

The hours glided on, and some time, perhaps soon after midnight, the skipper rose again from where he had lain apparently asleep, but really watching the speakers attentively, and coming aft this time with one of his men, the sailor at the helm was changed, and the other went forward to throw himself down as if to sleep.

"Will not the effendi lie down and take his rest now?" said the skipper to Yussuf. "The day will not be very long before it comes, and then it is no longer time to sleep."

Yussuf quietly repeated the man's words to the professor, who replied coldly:

"Tell the Greek captain that he is paid to convey us to our journey's end, and that it is not for him to presume to interfere as to the way in which we pass our time. Tell him we know the night from the day."

Yussuf interpreted the words, and the Greek smiled and replied in the most humble manner that perhaps the English excellency did not know how bad it was for strangers to expose themselves to the night air. That he was anxious about them, and wished them to go into the little cabin to be safe.

"Tell him to mind his own business," said the professor shortly, and this being interpreted the man slunk forward, and the professor said softly:

"There is no doubt about it, Yussuf; the man is a scoundrel and has bad intentions."

"He is a pig," said the Muslim in a low voice full of contempt; "but he and his men will be afraid to show their teeth to your excellencies if we are watchful and take care."

Towards morning the man came aft again, but he did not speak, and just at sunrise Lawrence awoke to come hurriedly out of the cabin where Mr. Burne was still sleeping.

"I thought you would have called us," he said; "I thought we were to watch."

"So you are," said the professor smiling. "How have you slept?"

"Oh, deliciously—all the night. I never do at home, but lie awake for hours."

"Even in a comfortable bed!"

"Even in a comfortable bed," replied Lawrence. "But you must be very tired. I'll call Mr. Burne now."

"No, let him lie. He is a bit of an invalid too. Suppose you go and have a sleep now, Yussuf; my friend here and I will watch."

The Turk smiled.

"Your servant once went without sleep for six nights in a time of danger. He slept a little upon his horse sometimes. One night without sleep! What is it? A nothing. No, your excellency must not ask me to sleep now. A short time and we shall be ashore, and away from these Greek dogs, who think we are without arms; then thy servant will lie down and sleep for hours. Last night, to-night I shall not sleep."

The bright morning, the glancing sea, and the soft breeze seemed to take away all the fancies and suspicions of the night. The shore was in sight—the mainland or one of the beautiful Grecian isles, and to make matters more pleasant still Mr. Burne was in the most amiable of tempers.

"I must have been out of order when we were crossing the Channel," he said smiling. "I thought it was sea-sickness, but it could not have been, for I am as well as can be out here in this little boat."

The professor was almost annoyed with himself for his suspicions about the Greek and his men, for an easier, happier-looking set it would have been impossible to find. They smiled and showed their teeth, as they lounged in the front of the boat or took their turn at the helm, and then picked out some sunny spot where the tall sails cast no shade and slept hour after hour. When they were not smiling or sleeping, they were eating melon, bread, grapes or olives, or watching like dogs to see if any food was going to be given them by the travellers.

The sail was glorious, and at first great way was made, but in the course of the afternoon the wind dropped, and the little vessel hardly moved through the water.

"This is vexatious," the professor said. "I am anxious to get to our journey's end."

"Don't say that," said Lawrence, almost reproachfully; "one seems to be so happy, and everything is so delightful out here in the sunshine. I should like to go sailing on like this for ever."

"If we had some cushions," put in Mr. Burne, who

had overheard his remark. " Well, it doesn't matter to a few days, one way or the other, Preston," he continued; " we are very comfortable considering, my back's better, and this is easy travelling, so never mind about Yussuf's suspicions. All nonsense."

That day glided away, the brilliant night came, and with it the nervous feeling of all being not as it should be.

Nothing more had been said to Mr. Burne till quite evening, but then the professor felt it to be his duty to speak of the suspicion, and did so; but the old lawyer laughed.

" What nonsense, Preston!" he said; " why, the man and his crew are like so many good-tempered gypsy boys. No, sir, I am not going to be scared because the night is coming on. Poor fellows, they are honest enough. That sour Turk—I don't like the fellow—has been filling our heads with nonsense to make himself seem more important. It's all right."

" I hope it is," said the professor to himself, and in due course he lay down, but not to sleep.

During the day, by a quiet understanding, he and Yussuf had taken it in turns to snatch an hour's repose, with the result that they were far better prepared to encounter the night than might have been supposed.

" We will lie down, excellency," Yussuf took the opportunity of whispering; " but one of us must not sleep."

After a time the old lawyer, who had been leaning back watching the stars from far above till they seemed to dip down in the transparent sea, yawned aloud, and then began to talk in an unknown tongue, using a strange guttural language which for the most part consisted of a repetition, at regular intervals, of the word "*Snorruk*," and this had a wonderful effect upon his companions, who had felt listless and drowsy after the hot day; but the coolness of the night and the interesting nature of Mr. Burne's discourse effectually banished sleep, and hence it was that, when the skipper and a couple of his men came stealing aft to apparently change the steersman, the professor sat up, and Lawrence saw that Yussuf was wide awake and on the *qui vive*.

This occurred three times, and then the rosy morning lit up the tops of the distant mountains, and made the sea flash as if it were all so much molten topaz.

A pleasant listless day followed, and another and another, during which the travellers slept in turn, and watched the various islands seem to rise out of the sea, grow larger, and then, after they were passed, sink down again into the soft blue water.

It was a delicious dreamy time, the only drawbacks being the suspicions of the boatmen, and the cramped nature of the space at disposal.

They sailed on and on now, with the water surging beneath their bows and the little vessel careening over in the brilliant sunshine; but they were still far from their destination, and now the question had arisen whether it would not be wise to put in at the principal port of Cyprus, which they were now nearing, to obtain more provisions, as the wind was so light that the prospect of their reaching Ansina that night was very doubtful.

The evening had come on, with the sun going down

in the midst of a peculiar bank of clouds that would have looked threatening to experienced eyes; but to the travellers it was one scene of glory, the edges of the vapours being of a glowing orange, while the sky and sea were gorgeous with tints that were almost painful in their dazzling sheen. There was not a breath of wind, not a sound upon the smooth sea. The sails hung motionless, and the heat was as oppressive as if those on board were facing some mighty furnace.

"Very, very grand!" said Mr. Burne at last, after he had sat with the others for some time silently watching the glorious sight; "but to my mind there's too much of it. I should like to have it spread over months, a little bit every night, not like this, all at once."

"Oh, Mr. Burne!" cried Lawrence reproachfully.

"I once saw a pantomime many years ago, when I took some of my sister's children to a box I was foolish enough to pay for. This reminds me of one of the scenes, only there are no sham fairies and stupid people bobbing about and standing on one leg. Just when everything was at the brightest a great dark curtain came down, and it was all over, and it seems to be coming here, only it's coming up instead of coming down. Heigho—ha—hum! how sleepy I am!"

He lay down as he spoke close under the low bulwark, and as he did so Lawrence glanced forward and saw that the gorgeous sunset had no charms for the sailors, for they were lying among the baskets fast asleep, their faces upon their arms, while, upon looking aft, the man at the helm was crouched up all of a heap sleeping heavily.

"It is very beautiful," said the professor; "but I daresay some of our English sunsets are nearly as bright, only we do not notice them, being either shut up or too busy to look."

"Doesn't this curious stuffy feeling of heat make you feel drowsy, Mr. Preston?" said Lawrence, after a few minutes' silence, "or do I feel it because I am weak with being ill so long?"

"My dear boy," replied the professor laughing, "at the present moment I feel as if all my bones had been dissolved into so much gristle. It is the heat, my lad, the heat."

Lawrence lay back upon the deck with his head resting upon a pillow formed out of a doubled-up coat. He had tried going below, but the little cabin was suffocating. It was as if the bulkheads and deck had imbibed the sun's heat all day and were now slowly giving it out. To sleep there would have been impossible, and he had returned on deck bathed in perspiration to try and get a breath of air.

As he lay there he could see the old lawyer sleeping heavily, the professor with his head resting upon his hand, and his face glorified by the reflection from sea and sky, and their guide Yussuf seated cross-legged smoking placidly at his water-pipe, his dark eyes seeming to glow like hot coals.

Beyond him lay the Greek and his men upon their faces, motionless as the man at the helm, and then all at once the muttering bubbling noise made by Yussuf's pipe seemed to be coming from the old lawyer's parted lips, and the pipe, instead of justifying its name of "hubble-bubble," kept on saying *snorruk—snorruk*, after the fashion of Mr. Burne. Finally, there was nothing —nothing at all but sleep, deep, heavy, satisfying sleep that might have lasted one hour, two hours, any length of time. It seemed as if there was no dreaming, till all at once Lawrence imagined that the professor was bitterly angry with him for getting better that he jumped up and kicked him violently, and that then, as he tried to rise, he stamped upon him, and the stamp made a loud report. He was awake.

Awake, but in a dazed, puzzled state, for all was pitchy dark, and as he jumped up he was knocked down again, and would have gone over the side had he not struck against and clung to one of the ropes which supported the mast.

About him a terrible struggle was going on; there was heavy, hoarse breathing; men were trampling here and there with falls and struggles upon the scrap of a deck.

Then Lawrence turned cold, for there was a yell and a splash, followed directly after by a blinding flash of light and a loud report.

The struggle went on for a few moments longer, seemed to cease, and a voice that he recognized said some words hastily in Greek, which were replied to in hoarse panting tones.

Then the professor's welcome voice arose out of the pitchy darkness.

"Lawrence! Lawrence! where are you?"

Before an answer could be given there was the dull thud of a heavy blow, and the professor roared more than spoke the one word:

"Coward!"

The struggle was resumed for a moment or two, while the Greek skipper yelled out some order; but before it could be executed there came from out of the darkness a sharp hiss and a loud roar. Lawrence felt himself drenched by what seemed to be a cutting tempest of rain, and then it was as if some huge elastic mass had struck the boat, capsizing it in an instant. The lad felt that he was beneath the surface of the water, the sudden plunge clearing his faculties and making him strike for the surface.

As he rose he had touched a rope, which he caught at with the instinctive clutch of a drowning man, and found that it was attached to something which enabled him to keep his head above the water, but how it was or what it all meant he could not comprehend in the midst of the deafening rushing noise of the wind and the beating stinging blows of the surf that was flying over him.

All at once from out of the darkness a hand seemed to be stretched forth and to grasp him by the collar of the light Norfolk jacket he wore.

In spite of himself he uttered a cry of horror, but the grasp was not inimical, for he felt that he was drawn up on to what seemed to be a heaving piece of wood-work, and then a strong arm was passed round him, a man's breast pressed him down, and the rush and roar and confusion increased.

There were times when he could scarcely breathe, the wind and spray stifling him till he could turn by an effort a little aside. Then for long periods together, as they seemed, they were under water, as some wave leaped over them. In fact, after a few such experiences

he was half insensible, and every struggle towards recovery was met by a new attack.

How long it lasted the lad never knew; all he could
comprehend was that he was floating upon something
in the midst of a wildly tempestuous sea, and that the
wind and spray seemed to have combined to tear him
from where his feeble efforts were aided by a strong
man's arm.

Once or twice he fancied he heard a shout, but he
could not be sure, and he could make no effort to
understand his position, for the storm that had stricken
the boat so suddenly robbed him more and more of the
power to move.

It was like another waking from sleep, to feel that
his head was being raised a little more from where it
drooped, and someone pressed a pair of lips to his ear
and spoke.

He could not answer, he could not even move, for
though the voice was familiar, its import did not reach
his brain, and he lay perfectly inert till it seemed as if
the sea and wind were not beating so hard upon his
face, and that he could breathe more easily.

Then it was not so dark, for the stars were coming
out, and he found himself gazing at a great black veil
that was being drawn over the heavens.

The next thing he heard was a voice, a familiar
voice, speaking, and another which he recognized, and
which came from close by, answered, but what was said
he could not tell.

There was another confused half-dreamy time, and
then it was comparatively light. The spray had ceased
to beat, and the mass of wood upon which he had been

dragged was rising and falling in a regular drowsy rocking fashion, while now he felt bitterly cold.

"I cannot get to you, Yussuf," said the familiar voice again. "If I attempt to move he will slip off into the water. Safe?"

"He is alive!" came in a low deep voice from close by Lawrence's ear, and then there was a fierce puff of wind again, and with it the dreamy sensation once more.

CHAPTER XI.

CAST ASHORE.

HEN Lawrence came to himself again there was more vigour in his brain, and he was conscious that he was on the side of the boat held fast by Yussuf. The wind was blowing fiercely, and had seized hold of a portion of a half-submerged sail which had filled out into a half sphere, and they were going swiftly through the water.

The stars were shining brightly; there was no more spray, and as he recovered himself he could see, right at the far end of the boat, the dimly defined head and shoulders of the professor, whom he knew by his great beard, and he seemed to be supporting Mr. Burne.

Between them, seated high and clear of the water, were the Greek skipper and a couple of his men, holding on tightly in a bent position.

There was deep silence now, save the ripple made by

the boat in going through the water, which it did at a
fairly rapid rate, seeing how it was submerged; but the
wind having filled the portion of the sail, seemed to be
raising it more and more from where it lay in the water,
and as a natural consequence the more surface was
raised and filled, the more rapidly the other loose
portion was dragged up, distended, and drew the boat
along.

For a full hour no one spoke. The travellers were
divided by the Greek and his men, who held the post
of vantage, and there was a growing feeling in every
breast that if any attempt were made to get into a
better position, the enemy would be roused to action,
and perhaps thrust them from their precarious hold
into the sea.

By degrees Lawrence began to get a clear understanding of what had happened, and as far as he could
make out the suspicions of Yussuf had been quite
correct. The Greek and his men, for purposes of
robbery, had made an attack during the night when all
were asleep, and in the midst of the struggle one of
the terrible squalls, whose threatenings they had not
read on the previous evening, had suddenly struck and
capsized the boat, to which they were now desperately
clinging for life.

Lawrence felt too much numbed to speak to Yussuf,
or even to shout to the other end of the boat, where the
professor was clinging, and his companion was too intent upon holding him in his position to care to make
any remarks.

The breeze blew very coldly, and the lad knew that
if it increased to any great extent, and the waves

rose, they must all be swept off; but the wind showed
more disposition to lull than increase, the sail flapping
and sinking once, but only to fill again and bear them
steadily on. For the squall had exhausted its violence;
the intense heat had passed, and the sea rapidly grew
more placid as they were borne along.

There was something strange and terrible, and
sufficient to appal a heart stronger than that of a boy
who had suffered from a long and severe illness. The
darkness seemed to float as it were in a thick transparent body upon the surface of the sea, while far
above the stars shone out clearly and spangled the sky
with points of gold.

Where were they being borne? What was to be the
end of it all? Were they to cling there for an hour—
two hours, and then slip off into the sea?

It was very terrible, and as he grew cold, a strange
sensation of reckless despair began to oppress Lawrence, mingled with a feeling that perhaps after all it
would be better to let go and slide off the boat so as to
arrive at the end.

These despondent thoughts were ended upon the
instant by a movement made by one of the Greeks
who were crouching in the middle of the boat.

He seemed to be quitting his position slowly and to
be creeping towards where Yussuf was clinging.

At that moment the Turk heaved himself up; there
was a quick movement of his arm; and Lawrence clung
spasmodically to the boat, for he felt himself slipping.

In his agony he did not hear the click made by the
pistol the guide had snatched out and held before him;
neither could he understand the Turk's words, but

they were full of menace and evidently embodied a threat.

The Greek uttered an angry snarl and snatched a knife from his waist, as he crept on and said something, to which Yussuf replied by drawing trigger.

The result was a click, and the Greek laughed and came on; but just as he was nearly within striking distance Yussuf drew trigger again, and this time there was the sharp flash and report of the pistol, while for a moment the smoke hid the man from view, but a cry of agony and fear was heard.

The breeze cleared the smoke away directly, and revealed in the dim starlight the form of the Greek lying back and one of his companions crawling to his side.

The Turk uttered a few words full of warning, and the second Greek paused to speak in a low pleading tone, to which Yussuf responded by lowering his arm and watching his enemies while one helped the other back to his place where he had clung.

" Is he much hurt?" came from the other end of the boat.

" I cannot say, excellency," was Yussuf's reply in English. But directly after he roared out a few words in Greek, with the pistol pointed; for as soon as the wounded man was crouching in the central part of the boat he said something fiercely, and his two followers began to creep towards where the professor and the old lawyer clung.

It was plain enough to all what Yussuf had shouted, with pistol aimed, for the two Greek sailors cowered down as if seeking to shelter themselves behind their wounded skipper, and for a space no one moved or spoke.

Yussuf was the next to break the silence with a few words of warning which made the Greeks creep back to their old position, and then what seemed to be a terrible space of time ensued in the darkness that grew colder and colder, and where it seemed to be vain to look around for help. No one moved or spoke, but all were animated by the same intense longing, and that was for the light of day.

Morning seemed as if it would never come. Right in front there was a great black cloud touching the sea and rising high; but though the wind set towards the cloud, which grew higher and broader, they knew that at any time the breeze might change to a furious squall, coming from where that cloud was gathering; and when it came it would be to find them numbed and cold, and unable to resist its violence and the beating waves.

The helpless drowsy sensation was attacking Lawrence again, and he would have slipped back into the sea but for the strong arm about him. The dimly-seen figures grew unreal and as if part of a dream, and he was falling more and more into a state of unconsciousness, when, as if by magic, there was a patch of light in the sky before them, to right of the great cloud; there was a dull murmur ahead; then more light, and, as if by some rapid scenic effect, the stars paled, the sky grew gray, then pink, red, glowing orange, and it was morning.

Yussuf uttered a low cry of joy, for the dark cloud ahead of them was a high mountainous land, whose

topmost points were beginning to blush with the first
touches of the sun that was rising directly behind.

"We are safe, excellencies!" cried the guide. "In an
hour this wind will carry us to the shore."

"The boy!" cried the professor in a low voice that
told of exhaustion.

"He is here and safe," was the reply. "It is day
once more, and we can perhaps better our position."

The words were hopeful and had a stimulating effect,
but nothing could be done. The Greeks could not be
trusted, even under the influence of threats, to go to
the help of the professor; and Yussuf dared not quit
his own charge, for Lawrence was too much exhausted
to be left alone; so there was but the one hope—to
wait and remain clinging to the side of the boat until
the breeze carried them ashore.

As the sun rose warm and bright it brought with it
hope and sent a glow through the chilled forms of all,
but the morning light made nothing else clear. They
were just as they had made themselves out to be in
the darkness.

The sail had been filled now till it was of a goodly
size, and they were borne more swiftly still towards
what seemed to be a barren rocky coast; but the same
dread was in the heart of each of the travellers, and
that was lest when the sun rose higher the power of
the wind should fail, and, slight as the currents were
in that part of the world, they might be swept past
the land unseen.

The dread was needless, for at the end of about a
couple of hours of the most intense anxiety the boat
was blown close in to the beach, and struck with a

THE PROFESSOR AND YUSSUF BRING LAWRENCE ASHORE.

bump that changed her position, shaking Yussuf and his companion from their hold.

But it was into the shallow transparent water, and, gaining his feet, Yussuf tried to raise Lawrence on to his shoulders; but he was so stiff and cramped that he could only hold the lad beneath his arm and wade with him ashore.

CHAPTER XII.

A WARM LAND WELCOME.

THE distance was only some forty yards, and Yussuf was quite half-way there when he was met by the professor, who came staggering down to his aid, and between them they carried Lawrence the rest of the way, to lay him beside Mr. Burne in the full sunlight and upon the soft warm sand.

The three Greeks were already ashore selecting a spot a good hundred yards away, and they could be seen to be stripping the clothes from their wounded captain, and then one of them appeared to be binding a cloth round his leg, showing where Yussuf's bullet had taken effect.

By way of precaution Yussuf's first act was to take out his pistol, and swing it about to get rid of all the water possible before uncharging it, and laying it with its cartridges in the sun to dry, in the hope that some of them might prove to be uninjured, the water not

having been able to penetrate to the powder, though
it was extremely doubtful.

His next act was to take out his pipe from a pocket
in his loose robe, and place that with his bag of tobacco
and little tinder-box and matches also in the sun, which
was rapidly gaining power, all of which being done he
proceeded coolly enough to slip off his garments, to
wring them and spread them upon the glowing sand.

Meanwhile the professor was dividing himself
between Lawrence and the lawyer, then lying in the
warm sunshine, whose influence rapidly made itself
felt, and seemed to carry strength as well as a pleasant
glow.

"Well, Lawrence," said the professor anxiously,
"how do you feel?"

"Not quite so cold," was the reply, "but very stiff
and hungry."

"Hah!" ejaculated the professor, "then you are not
very bad.   Can you follow Yussuf's example?"

Lawrence hesitated.

"Take my advice, my lad.   Take off and wring your
clothes as well as you can, and then, in spite of being
soaked with the sea-water, go down and have a quick
plunge, and then walk or run about till you are dry."

The advice seemed so droll, that now the danger was
past the lad laughed, but he saw that Yussuf was doing
precisely what the professor advised, and, weakly and
shivering a good deal, he did the same.

Freed by the evident lack of anything to apprehend
about the lad for the present, the professor turned to
Mr. Burne, whom he had been helping for some hours
to cling to the boat, and had sustained with a few

whispered words of encouragement in his feeblest moments.

The old man was lying in the sunshine just as he had sunk down upon his back, apparently too much exhausted to move, but as the professor went down on one knee by his side he opened his eyes.

"Not dead yet, Preston," he said smiling. "I say, don't laugh at me."

"Laugh at you, my dear sir?"

"For being such an old goose as to come upon such a journey. Oh, my back!"

"Come, come, it was an accident."

"Accident, eh? I say, we'll prosecute those murdering thieves of Greeks for this."

"One of them has met his punishment already," said the professor, "and Yussuf has severely wounded another."

"Yes. I was pretty well done then, but I saw him shoot that scoundrel. I believe the heathen dog was going to shove us off."

"There is no doubt about that," said the professor.

"But Yussuf? don't you think he was in league with the murderous rascals?"

"Yussuf? My dear sir!"

"Humph! No! He couldn't have been, could he, or he wouldn't have fought for us as he did at first, and then shot that scoundrel yonder? I hope his bandage will come off and he'll bleed to death."

"No, you do not," said the professor.

"Oh, yes, I do—a dog!" .

"No, you do not; and as to Yussuf—well, I need not defend him."

"Well, I suppose not. Boy seems to be all right,
don't he?"

"Yes, I think so. This warm sunshine is a blessing."

"Hah, yes, but I'm so stiff and sore I cannot move.
Preston, my dear boy, would you mind putting your
hand into my pocket and taking out my snuff-box.
I suppose it's all paste, but a bit of that would be, like
your sunshine, a blessing. It's all very well, but I'd
rather have a fire, a towel, a warm bath, and some
dry clothes. Hah, yes! Thank you. Now for some
paste."

He thrust the little box in and out among the dry
sand till the moisture was all gone, and doing this
dried and warmed his hands as well before he proceeded to open the lid, when he uttered a cry of satisfaction.

"Bravo, Preston! Dry as dust. Have a pinch, my
dear sir?"

"Thanks. No. I am drying a cigar here for my
refreshment, in the hot sand. I daresay my matches
are all right in their metal box."

"Just as you like. Smoking is all very well, but
nothing like a pinch.

"I am most anxious about the boy," said the professor.

"Must teach him to take snuff. Well, where are
we? Is this a desolate island, and are we going to be
so many Robinson Crusoes for the rest of our days?"

"Desolate enough just here," replied the professor;
"but it must be inhabited. It strikes me that we have
reached Cyprus."

"Then, my dear fellow, just look about, or shout, or do something to make the inhabitants bring me a bottle of Cyprus wine. Hah! a pinch of snuff is a blessing, and, bless me, how wet my handkerchief is!" he cried, as he struggled to his feet and took out and wrung the article in question before making the rocks echo with a tremendous blow.

"How do you feel?" said the professor.

"Bad, sir; but I'm not going to grumble till we get all right again. I must try and walk about to get some warmth into me. How beautiful and warm this sand is! Hah!"

He seemed to revel in the beautiful dry sand of the shore, which, with the sunshine, sent a glow into the perishing limbs of all, and to such an extent that in about an hour the sufferers were not so very much the worse for their adventure. The professor and Mr. Burne had lit cigars; Yussuf was enjoying his pipe; and Lawrence alone was without anything to soothe his hunger.

The wounded Greek lay at a distance where his companions had left him. The professor had been to him with Lawrence, and seen to his injury, the others paying no heed, and the injured man himself only looking sulky, and as if he would like to use his knife, even though he was being tended by a man who knew something of what was necessary to be done.

He was left then, and the professor and Lawrence joined Mr. Burne, who was very cheerful though evidently in pain.

"I say," he said, "those fellows had planned that attack."

" Decidedly," said the professor. " I feared it, though I did not say anything more to you."

" Then it was very ungentlemanly of you, sir," cried the old lawyer testily. " Lucky for you I was awake, sir, or we should all have been killed in our sleep."

" I thought you were fast asleep, as, I am ashamed to say, I was."

" Oh, you own you were, professor."

" Fast."

" Then I'll own I was too. It seems, then, that Yussuf was on the watch and met them."

" Exactly so, and saved our lives."

" Well, I don't know about that, but he certainly kept the boy from drowning during the night, for I couldn't stir to help him. I don't dislike that fellow half so much as I did; but I wish to goodness he could do as these Turks and Persians did in the *Arabian Nights.*"

" What's that?" said the professor.

" Conjure a breakfast up for that poor boy."

" It strikes me," said the professor, who was watching where Yussuf had posted himself on the edge of the sea, " that that is the very thing he is about to do."

" Eh? what do you mean?"

" Oh, I say, Mr. Preston, don't talk about food if there is none," cried Lawrence, " for I am so hungry."

" I mean this," said the professor, " that the two Greeks down there are evidently trying to get something out of the boat, and if they find anything to eat, Yussuf is there with his loaded pistol, and he will certainly have a share."

In effect the two sailors had stripped, and were busy

in the shallow water doing something, and in a short time they had contrived to thrust the boat out, and, by using the masts as levers, completely turned her round, so that her deck was parallel with the shore.

The men were evidently working hard, and in a short time they had got the vessel so closely in that they were able to lower the sails, or rather run them down to the foot of each mast, with the result that, by the help of hard work with a spar they partly raised the side of the boat that was submerged, its natural inclination to resume its normal position aiding them; and at last, after several attempts, they succeeded in getting at one of the baskets of provisions that had fortunately not been washed away.

As they dragged this out and waded ashore, they were for making off in the direction of the spot where their wounded skipper lay, but a few sharp orders from Yussuf stopped them.

They were not disposed to yield up their prize peaceably, for each man's hand went to his knife, and the professor ran down to Yussuf's help.

But there was no need. The Turk went close up to them, pistol in hand, and the men stooped and lifted the basket, carrying it between them sulkily to where Mr. Burne and Lawrence were breathlessly watching the proceedings.

The water streamed and dripped from the basket as they bore it over the sands, and plumped it down, scowling fiercely, where they were told to stop. Then turning, they were going off, but the professor bade them stay.

They did not understand his words, but their tone

was sufficient command; besides there was Yussuf's
pistol, which acted like a magician's wand in ensuring
obedience.

"Tell the scoundrels that we will behave better to
them than they have to us, Yussuf," said the professor;
and he took out from the dripping basket a great
sausage, a bottle of wine, and one of the tins of biscuit
that were within.

"Am I to give them this food, effendi?" said Yussuf
calmly. "You will get no gratitude, and the dogs
will murder us if they get a chance."

"Yes; give it to them," replied the professor. "Coals
of fire upon their head, O follower of Mahomet.
There, bid them eat. We may want to make them
work for us."

Yussuf bowed, and handed the food and wine to the
two Greeks, who took what was given them without a
word, and went to join their companion.

CHAPTER XIII.

HOW TO BALE A BOAT.

"AH!" ejaculated Mr. Burne, after they had
made a hearty meal, seated upon the warm
sands. "I don't know that I like my biscuit sopped, and there was more salt than
I cared for, but really I don't feel as if I had done so
very badly. Another taste of that wine, Preston.
Hah! well, we might have been worse off."

This was the general opinion, for matters looked better now, and a discussion arose as to what they were to do next; whether they were to travel along the coast till they came to some village, or, as Yussuf suggested, try to get the boat baled out and righted, and once more make for Ansina.

Yussuf declared that they were undoubtedly on the western coast of Cyprus, but he could not tell them how far they might have to journey, and it would be terrible work for Lawrence, who was too weak to walk far, so the Muslim's suggestion was received; and its wisdom was endorsed by the action of the Greeks, who had carried their skipper down to the boat and seated him upon the sands.

" We are three strong men against two now," Yussuf had said, "for we will not count the wounded master, or the young effendi here. The men shall empty the boat of water, and they shall take us across to the coast."

" But suppose another storm should come?" said Mr. Burne.

" If another storm should come we should meet it like men, effendi," said the Turk gravely. " That white squall last night saved our lives, for I was mastered."

" And so was I," said the professor. " You are right, Yussuf; but we must not let ourselves be surprised again. I had no business to sleep."

" We should not have been surprised if yon Greek dog had not struck me down when he was pretending to be asleep by the helm. But see, effendi, he is ordering them to try and empty the boat. Let us go down and help."

The remains of the food were placed in the basket,

which was carried down and left in the sun to dry, not far from where the Greek skipper was seated, holding his wounded leg.

The tide there was very slight, but still it was falling, and this helped them in their plans.

The two Greeks were hard at work with the spar, using it as a lever; and twice over they obtained so good a purchase that they raised the submerged side just above the water, but it slipped back directly.

The professor did not hesitate, but said a few words to Yussuf, who handed his loaded pistol to Lawrence, tucked up his garment, and waded into the water at once along with Mr. Preston.

"Humph! just as they were getting so nice and dry," said Mr. Burne. "Well, when one is in Cyprus, one must act like a Cypriote, eh, Lawrence, my lad? I say, fancy one of my clients seeing me doing this."

He took off his coat, and rolled up his shirt-sleeves, nodding laughingly at Lawrence.

"Look here, my boy," he said, "if that Greek rascal there moves, just you go up and shoot him somewhere. Don't kill him, but we cannot stand any of his nonsense now."

The two Greek sailors stared as the three travellers came wading to them, and seemed disposed to leave off their task; but Yussuf gave them their orders direct from Mr. Preston, who made them get out some pieces of board and cut loose a couple of spars.

The result of this was that one of the long spars was securely lashed by their aid to the top of the principal mast which acted as a lever, when all took hold of the spar and pushed upwards. By this means the side of the boat was raised a foot or so, and could not sink back, for the free end of the spar rested on the sand. Then another foot was gained, the end of the spar being dragged along, and so on and on, till from being where it was lashed to the top of the mast, quite an obtuse angle of the widest, it was by degrees worked into a right angle, and by that time the submerged bulwark was quite out of the water, and the keel touched the bottom and kept them from moving the boat any farther.

The next thing to be done was to bale out the enormous quantity of water within, and there was no bucket or anything of the kind; but the professor was equal to the occasion. There was a small box in the big provision basket and the biscuit tin. These were emptied at once, and the two sailors set to work baling, while, as soon as it was possible, an attempt was made to get something serviceable out of the little cabin.

The search was vain, but just then one of the sailors took out his knife, left the biscuit tin with which he was baling, and going forward thrust down his knife-armed hand, and cut free a good-sized cask which was lashed there for the purpose of holding water.

This floated up directly, and when the man had got so far, he stood holding on and looking at it.

Yussuf had seized the biscuit tin, and was baling so as to lose no time, but the professor waded to the sailor, tossed the cask over, and following it, dragged it out on to the sandy shore, where the sea-water with which it was now filled ran gurgling out of the big bung-hole.

While it was emptying the professor walked some

little distance to where a few pieces of rock were lying, and securing one weighing about half a hundred-weight, he brought it back, set the cask up, and dashed in its head.

This made a baling implement of wonderful power, as soon as it was floated back and lifted into the boat. Certainly it took two men to use it, but the professor called to Yussuf to give the baling tin back to the Greek, and come to his side, and then Christian and Muslim set to work, stripping to it and displaying energy that made the Greeks work the harder in spite of the burning sun. For seizing the cask, as he stood waist-deep, the professor depressed and sank it, and as soon as it was full, he and Yussuf raised it between them till the edge was against the low side of the boat, and then they tilted it, sending its contents into the sea.

It was slow and terribly laborious work, but at the end of an hour the amount they had discharged was something tremendous, and after a rest for refreshment, the baling went on till, towards evening, the felucca was afloat once more, and riding to a little anchor cast out upon the shore.

There was still a great deal more water in her; but everyone was wearied out, and the professor gave the word for a cessation of labour, when some more provision was secured, with wine, and fairly distributed, when the Greeks encamped by their skipper, and the travellers went up close to the rocks, where a little thread of delicious fresh water trickled down and lost itself in the sand.

This was a treasure to the travellers, and at the

professor's desire Yussuf filled the biscuit tin, and took it to the Greeks, who, however, only laughed and said they preferred the wine.

The deliciously warm evening was spent in drying the wet garments in the heated sand, and in resting. The professor, who seemed a good deal fagged by his exertions, would hardly hear of sleeping, but was exceedingly anxious about Lawrence, who, however, seemed to be none the worse for the past night's exposure, the warmth of the day and the rest he had had having recouped him to a wonderful extent. Mr. Burne, too, though he had worked very hard, declared that he never felt better, and after smoking a cigar, which he took as a sandwich between two layers of snuff, preparations were made for the night, it being decided to lie down early and rise at daybreak, when a couple more hours' work would, it was considered, make the felucca in a condition to sail at any time.

The professor insisted upon Yussuf lying down at once to get the first rest, so as to be roused up towards midnight to take the watch.

He consented rather unwillingly, and then the point had to be settled who should have the pistol and take the first watch.

The professor wished to commence, but Mr. Burne was so indignant and insisted so sternly that the pistol was handed to him, after Yussuf had been asleep for about a couple of hours, and then Mr. Preston and Lawrence sought their sandy couches, and lay for a little while listening to the soft murmur of the sea, and watching the brilliant stars in the dark sky and in the purply black water, while with regular and slow beat, like a sentry, Mr. Burne walked up and down, pistol in hand.

Lawrence lay awake long enough to hear the professor's deep breathing, and his muttering of something once or twice. Then he lay gazing at the old lawyer, thinking how comical it was, and what a change from Guilford Street in busy London, till it all seemed to be dim and strange and dreamlike.

Then it really was dreamlike, for, though the old lawyer was still marching up and down before Lawrence's mental vision, it seemed to him that he had swollen out to ten times his natural size, and that he was not walking to and fro between him and the sea, but in front of the railings in Bloomsbury, and that, to prevent his making a noise and disturbing the sleepers, he had wound list all about his boots, which now made not a sound upon the pavement.

To and fro, to and fro he seemed to go, till his head swelled and swelled and no longer appeared to be a head, but a great rough grenadier's cap, and it was no longer Mr. Burne, but one of the sentries in front of the British Museum, who marched, and marched, and marched, till he marched right out of sight, and all was blank as a deep, deep sleep is sometimes, from which the lad started into wakefulness just before dawn, upon hearing the professor say loudly:

"Eh? What? Is it time?"

CHAPTER XIV.

HOW MR. BURNE KEPT WATCH.

ES, effendi, quite time," said a stern voice which Lawrence, as he sat up, recognized as Yussuf's; and there was the grave-looking Turk, misty and strange of aspect, bending down.

"Quite time, eh?" said Mr. Preston yawning.

"Quite time, effendi Look there!"

Mr. Preston rose and gazed in the direction of the Turk's pointing finger, which was directed towards something indistinctly seen a few yards away.

"Mr. Burne! Asleep!" said the professor quickly.

"Yes, effendi; I lay down to rest as you bade me, and I slept, expecting to be called later on to watch; but I was not awakened, and slept heavily. I was weary."

"But Mr. Burne was to watch for only three hours as near as he could guess, and then call me. It is too bad. Those scoundrels might have stolen upon us in our sleep."

Lawrence had risen and joined them.

"Poor fellow!" he said softly; "he must have been tired out. Let me watch now, Mr. Preston."

"No," said the professor sternly. "Lie down and sleep, my lad. Sleep brings strength. You shall have your turn as soon as you are well enough."

"Thy servant will watch now," said Yussuf. "It is nearly day."

"It is too bad," said the professor again; and with the Turk he walked to where Mr. Burne lay fast asleep—so soundly, indeed, that he did not stir when Yussuf bent down and took the pistol from his hand.

"Let him sleep, then," said Mr. Preston rather bitterly. "I will watch;" and as he spoke he looked in the direction of the Greeks' camp.

"Let thy servant," said Yussuf quietly; "I am well rested now."

The result was that Lawrence, after a glance round to see that everywhere it was dark and still, once more lay down to sleep, leaving Mr. Preston and the Turk talking in a low voice about their proceedings the next day.

Then once more all was blank, but to the lad he did not seem to have been asleep a minute when he heard voices and started up, to see that it was broad daylight, and that Mr. Preston and Yussuf were in earnest conversation with Mr. Burne, who was sitting up rubbing his eyes.

"Been asleep!" he cried; "nonsense! I don't believe I have closed my eyes."

"No," said Mr. Preston as Lawrence hurried up. "I do not suppose you did. It was nature, and she laid you down comfortably on this soft sandy bed."

"But you astound me," cried the old lawyer. "I can't believe it."

"Quite true all the same," said the professor; "but never mind now."

"It is of no use to mind, my dear sir. We must make the best of it."

"Of course, but you should have awakened me when you felt weary."

"Yes, exactly; I meant to—I—dear me! I remember now. I thought I would lie down for a few moments to take off a drowsy feeling. I meant to get up again directly, strong and refreshed. Dear, dear, dear! I am very sorry! So unbusiness-like of me! What time is it?"

The professor smiled.

"About four, I think."

"Ah, yes; it must be about four," said the old lawyer looking about him and encountering the stern eyes of Yussuf, which were full of reproach. "Good job the Greeks did not come and disturb us."

"They did not disturb you, then?" said the professor gravely.

"No; not they—the scoundrels! They had too serious a lesson in the boat, and—"

He stopped short and looked in the direction of the spot where the three Greek sailors had lain down to sleep the night before, and then he turned his gaze out to sea.

"Why, where are they?" he exclaimed at last.

"Where, indeed!" replied the professor.

"You don't mean to say—you don't want to make me believe that they are gone!" cried Burne excitedly.

"They are not anywhere near here on shore," replied the professor; "and the boat has sailed away. There is only one in sight, miles away yonder. That may be it, but I am not sure."

"Do you mean to say that those scoundrels have

taken advantage of our being asleep to get on board the boat and escape?" said the lawyer angrily.

"That is the only point at which I can arrive," said the professor. "Look around and judge for yourself."

The old lawyer looked sharply about him and then walked slowly away.

"A mistake—a mistake," he muttered; "I ought never to have come upon such a trip. Not fit for it—not fit for it. Disgraceful—disgraceful! I never—never could have believed it of myself."

He stopped and turned back.

"Send away this man," he said quickly.

Yussuf turned and walked away without another word.

"Preston," exclaimed the old lawyer, "I don't know what to say in my defence. I have nothing to say, only that I never felt anything so bitterly before."

"Then say nothing," replied Mr. Preston coldly. "You were overcome by sleep, and no wonder. But it was a terrible risk to run. Fortunately these men were cowed by what had previously taken place, and they could not know but what we were keeping a good watch."

"It is inexcusable," cried Mr. Burne. "I feel as if I could hardly look you in the face again. Left helpless here! For goodness' sake, Preston, tell me what we are to do."

"Quietly consult together what is to be done," was the reply. "There, man! pray, don't look at me in that imploring way."

"But it is so inexcusable," cried Mr. Burne.

"Wait a bit," said the professor smiling; "my turn

may come soon, and I shall have to ask your pardon for doing wrong. There! perhaps it is for the best. If we had retained the scoundrels they might have been too much for us and played us some far worse trick."

Mr. Burne was about to speak again, but the professor arrested him and suggested a walk along the shore to the north-east; but it was finally decided to partake first of an early breakfast, then to pack together what was left of the food and start at once upon a journey that they hoped would soon lead them to a village or town.

After a visit to the shore, where the deep blue water came softly rippling upon the sand, they sat down to their frugal breakfast by the spring, carefully husbanding the supplies, and then with enough provision to keep them for about a couple of days, they started off, this provision being the only luggage they had to carry, what few things they possessed having been annexed by the Greeks, who seized upon them by way of payment for the trip, as of course they would not have dared to make any claim after what had occurred; and besides, it was not likely that the skipper would care to show himself at any port frequented by Englishmen for some time to come.

CHAPTER XV.

THE LAWYER'S APOLOGY.

OR some distance the way was along good firm sand, and they got over several miles before the heat became too much for Lawrence, who was glad to sit down under the shade of a low cliff facing the sea and nibble one of the biscuits that had been pretty well soaked with sea-water, and drink from a rivulet whose presence suggested the halt.

When the heat of the day had somewhat abated the journey was continued; and, at last, when the night was beginning to fall and arrangements had to be made for sleep, the outlook was very black, for they were in a very desert place, and, though Yussuf and the professor both climbed eminences from time to time, there was not a trace of human habitation, while their supply of food was growing very short.

"Never mind," said the professor cheerily. "Let's have a good night's rest. I don't think we need set a watch here, eh, Yussuf?"

"It is always better to do so, effendi," said the Muslim, in his quiet thoughtful manner; "there is a great ridge of rocks yonder in front, and who knows what may be on the other side."

"But no one has seen us come here; and if they had, we have not much to lose."

"Except the Turkish gold the two excellencies have in the belts round their waists," said Yussuf quietly.

Mr. Preston started at this, but said nothing then. Later on he found that his thoughts had been shared upon the subject, for, as they sat close up to a projecting cliff, Mr. Burne leaned towards him and whispered:

"Did you tell the guide that you had a lot of money in your cash-belt?"

"No. Did you?"

"No."

"It is very strange," said the professor.

"It is worse," was the reply; "but, look here, for goodness' sake don't go making me uncomfortable by hinting that Yussuf has designs against us."

"I am not going to," said the professor shortly. "I agree that it is strange that he should know it, but I am going to place absolute faith in Yussuf. If I am deceived in the man so much the worse for me."

"But he is an unspeakable Turk, Preston, and you are always reading what the Turks are."

"I am always reading what their wretched government is. As a race I believe the Turks are a particularly grave, gentlemanly race of men."

"I am sure," said Lawrence, "that Yussuf is doing all he can in our interest."

"Tchah! stuff, boy! what do you know about human nature?" cried Mr. Burne angrily. "We are out here in the desert at this man's mercy."

"But he fought for us and saved me from drowning."

"Of course he did, boy; he is paid to do it."

"Then why don't you trust him, sir?" said Lawrence, speaking out boldly.

"Because very likely he is doing all this to save us

for himself. Suppose he robs us and then runs away
to Tadmor in the wilderness, or some other outlandish
place, what can we do? There are no policemen here."

"Hush," said Mr. Preston; "here he is."

Yussuf came gravely stalking down from above
where he had been taking a fresh observation inland.

"I can see nothing, effendi," he said softly. "We
must sleep and see what another day brings forth."

"Yes," said Mr. Preston; "and we are all weary.
But, Yussuf."

"Effendi?"

"How did you know that my friend, here, and I
carried belts containing gold?"

The Muslim looked from one to the other sharply,
and it was plain that he read the suspicion in their
eyes, for his own flashed, and a stern aspect came over
his countenance.

It passed away directly and his face lit up with a
smile.

"Simply enough, excellencies," he said. "Mr. Burne,
here, is always feeling his waist to find out whether it
is quite safe, or lifting it up a little because it is heavy."

"I? What? No such thing, sir—no such thing,"
cried the old lawyer angrily.

"Well, I have seen you do so a great many times,"
said Mr. Preston laughing.

"And so have I, Mr. Burne," cried Lawrence, "often."

"I deny it, gentlemen, I deny it," he cried; and
sitting up he involuntarily placed his hands just above
his hips, and gave himself a hitch after the fashion of
a sailor.

The professor burst into a hearty laugh; Lawrence

roared; and Yussuf's face was so comically grave that Mr. Burne could not resist the infection, and laughed in turn.

"There," he exclaimed; "I suppose I do without knowing it, and I am so cautious, too."

"But come," said Mr. Preston, turning to Yussuf, "you have not seen me do this, I think."

"No, effendi, never; but when we were busy baling the water out of the boat for these dogs of Greeks to escape, your garments were wet and clung to you tightly, and the shape of the belt could be plainly seen."

"Of course it could," said the professor bluffly. "Why, Yussuf, I believe now in the story about the dervish who was asked if he met the camel, and told the owners all about it: the lame leg, the missing tooth, the load of rice on one side, the honey on the other, and all without seeing it."

"Nonsense!" said Mr. Burne testily, "how could he?"

"Why, my dear sir, you must have forgotten that old tale. By the light impression of one foot in the sand, by the herbage not being evenly cropped, and by the ants being busy with the fallen grain on one side, the flies, attracted by the honey, upon the other."

"Bah!" exclaimed the old lawyer. "Eastern tales are all gammon. I don't believe in the East at all."

"Nor in people being cast ashore in desert places and having encounters with Greek sailors. Nor in their having a faithful experienced Mussulman guide, who fought for them and strove his very best to get them out of their troubles, eh, Burne? Well, I do, and I'm very tired. Good-night, Yussuf. You are going to sleep, I suppose?"

"No, effendi," said the Turk. "I shall watch till the stars say it is two hours past midnight, and then I shall awaken you."

"Humph! Wrong again," cried Mr. Burne testily. "I always am wrong. What are you laughing at, sir?"

"At you, Mr. Burne. I beg your pardon, I couldn't help it," said Lawrence.

"Oh, I'll forgive you, boy. I'm glad to see you can laugh like that, instead of being regularly knocked up with our troubles. I begin to believe that you never have been ill, and were shamming so as to get a holiday."

"Do you, sir?" said Lawrence sadly.

"No, my boy. Good-night. Good-night, Yussuf," he added, and then he raised an echo by blowing his nose.

"Good-night, excellency," said the Turk, rather haughtily; and soon there was nothing to be heard but the sighing of the night wind and the low murmur of the rippling sea.

There was little to see, too, in the darkness, but the figures of the reclining sleepers, and that of the grave sentinel, who sat upon a big mass of stone, crouched in a heap and looking as if he were part of the rock, save when he changed his position a little to refill his pipe.

The night passed without any alarm. The professor was awakened about two and took Yussuf's place, and soon after daybreak the others were roused, and the residue of the provisions was opened out.

"Be easier to carry when eaten," said Mr. Preston laughing.

He looked serious directly, for there was a peculiarly sombre frown upon Yussuf's brow, which suggested that he was thinking over Mr. Burne's suspicions of the previous evening, and his rather unpleasant way.

"Look here, Burne," the professor whispered, as they sat together on the sand eating their spare meal, "I think, if I were you, I would make a bit of an apology to Yussuf. He is really a gentleman at heart, and has been accustomed to mix a great deal with Englishmen. He is a good deal hurt by our suspicions, and it is a pity for there to be any disunion in our little camp."

"Camp, indeed!" cried the old man testily; "pretty sort of a camp, without a tent in it. I shall be racked with rheumatism in all my old bones. I know I shall, after this wild-goose chase."

"Let's hope not," said the professor; "but you will make some advances to him, will you not?"

"You mind your own affairs, sir. Don't you teach me. My back's horrible this morning. Can't you wait a bit. I was going to make amends if you had left me alone."

"That's right," said the professor cheerily. "I want him to have a good opinion of Englishmen."

Lawrence watched eagerly for Mr. Burne's apology, but he did not speak till just as they were going to start, when he stepped aside behind a rock for a few minutes, and then came out again and walked up to Yussuf with something coiled up in his hand.

"Look here, Yussuf," he said. "You're a stronger man than I am, and used to the country. I wish you would buckle this round your waist—out of sight, of course."

As he spoke he held out his heavy cash-belt, which was thoroughly well padded with gold coin, and then threw it over the Turk's arm.

Yussuf looked at him intently, and a complete change came over the man's face as he shook his head and held the belt out for Mr. Burne to take again.

" No, excellency," he said, " I understand you. It is to show me that you trust me, but you doubt me still."

" No, I do not," cried Mr. Burne. " Nothing of the sort. You think I do, because I said ugly things yesterday. But that was my back."

" Your excellency's back?"

" Yes, my man; my back. It ached horribly. There, I do trust you. I should be a brute if I did not."

" I'll take your excellency's word, then," said Yussuf gravely. " I will not carry the belt."

" Nonsense, man, do. There, it was to make you believe in me; but all the same it does tire me terribly, and it frets me, just where I feel most tender from my fall. It would relieve me a great deal, and it would be safer with you than with me. Come, there's a good fellow; carry it for me. I beg you will."

The Turk shook his head, and stood holding out the belt, turning his eyes directly after to Mr. Preston and then upon Lawrence.

" Come," continued Mr. Burne, " you surely do not bear malice because a tired man who was in great pain said a few hasty words. The belt has really fretted and chafed me. I am ready to trust in your sincerity; will you not trust in mine?"

Yussuf's countenance lit up, and he caught Mr.

Burne's hand in his, and raised it to his lips hastily, after which he opened his loose robe and carefully buckled the money-belt within his inner garment.

"That's the way," cried Mr. Burne cheerily; and he looked happier and more relieved himself; "and look here, Yussuf, I'm a curious suspicious sort of fellow, who has had dealings with strange people all his life. I believe in you, I do indeed, and whenever you find me saying unpleasant things, you'll know my back's bad, and that I don't mean it. And now, for goodness' sake, let's get to some civilized place where we can have a cup of coffee and a glass of wine. Preston, old fellow, I'd give a sovereign now for a good well-cooked mutton-chop—I mean four sovereigns for four —one a-piece. I'm not a greedy man."

Lawrence went forward to Yussuf's side, and these two led the way, along by the purple sea, which was now flashing in the morning sun, and the delicious air made the travellers feel inspirited, and ready to forget all discomforts as they tramped on in search of a village, while, before they had gone far, Mr. Burne turned his dry face to the professor and said:

"Well, did that do?"

"My dear Burne," cried the professor, "I am just beginning to know you. It was admirable."

"Humph!" ejaculated the old lawyer, who then blew a sounding blast upon his nose. "I am beginning to think that a neater form of apology to a man—a foreign heretic sort of a man—was never offered."

"It could not have been better. What do you think, Lawrence?" he added as the latter halted to let his elders catch up, Yussuf going on alone.

(318)   I

"I don't know what you were talking about," he replied.

"Mr. Burne's apology. I say it was magnificent."

"So do I," exclaimed Lawrence. "Capital."

"Humph! Think so? Well, I suppose it was all right," said Mr. Burne. "But I say," he whispered, gazing after Yussuf who was striding away fifty yards ahead and leaving them behind, "do you really think that money will be all right?"

"I say, Mr. Burne," cried Lawrence laughing; "is your back beginning to ache already?"

The old lawyer stopped short, and turned upon the lad with a comical look, half mirth, half anger in his countenance.

"You impudent young dog," he cried. "I knew you were shamming, and not ill at all. My back, indeed! Well, yes. Come along. I suppose it was beginning to ache."

CHAPTER XVI.

THE STARTING-POINT.

R. BURNE showed no more distrust, though Yussuf was striding away faster and faster, at a rate that Lawrence's strength forbade him to attempt to emulate; but the reason soon became evident. He was making for an elevation about a mile away, and upon reaching it he toiled up to the top, and as soon as he had done so he turned and took off his fez and began to wave it in the air.

"He has found out something," said the professor.

"If it is a hotel where we can get a good breakfast he shall have my advice for nothing any time he likes to come and ask it," said Mr. Burne, rubbing his hands.

"In London?" said the professor.

"Anywhere, sir. There, that will do. Don't swing your arms about like that," he continued, addressing the guide, who was of course far out of hearing. "Anyone would think that because he was right on the top of a hill he had caught the wind-mill complaint."

The three travellers were almost as much excited as Yussuf, and hurried on, Lawrence forgetting his weakness in the interest of the moment, so that it was not long before they reached the top—hot, breathless, and panting with exertion.

Their guide pointed to what appeared to be a group of huts a long way off.

"Is that all?" grumbled the old lawyer. "I thought you had found a place where we could have a comfortable meal."

"There will be bread, and fruit, and a boat, excellency," said Yussuf quietly; "and these are what you want, are they not?"

"I suppose so," replied Mr. Burne, gazing forward at what now appeared to be a cluster of small houses by the sea-shore, backed by a dense grove of trees, while in front, and about a quarter of a mile from the sands, lay three small boats. "It is not a desert place then," he grumbled, as they all went on together. "How far is it to that cluster of hovels?"

"About two miles, excellency."

"About two miles, and before breakfast," muttered

the old fellow sourly; but he drew a long breath as if he were trying to master his disinclination, and then turning to Lawrence with a grim smile he cried, "Now, look here, cripple against invalid, I'll race you; fair walking, and Mr. Preston to be umpire. One—two—three—off."

It was a fair walk of about an hour before they entered the cluster of huts, each surrounded by a good-sized fruit garden, the people standing outside and staring hard at the strange visitors who came along the shore, one of whom plumped himself upon the edge of a boat that was drawn up on the sands, another throwing himself down, hot and panting with exertion, while the two who were left a little way behind strode up more leisurely, one of them to ask for refreshment and a resting-place out of the sun.

"There is no mistake about it, Lawrence," cried the professor eagerly, "you couldn't have done that in England."

Lawrence laughed.

"But I am completely tired out," he exclaimed, wiping his face. "I could not have gone any further."

"Neither could I," groaned Mr. Burne. "Oh, my back, my back! Who won, Preston?"

"A dead heat, decidedly," said the professor laughing; but he was watching Lawrence the while very attentively, and asking himself whether he was letting the lad over-exert himself.

One thing, however, was plain enough, and that was that the sick lad had been allowed to droop and mope in his ailment. The serious disease was there, of course, but he had been nursed up and coddled to a terrible extent, and this had made him far worse than he would have been had he led an active country life, or been induced to exert himself a little instead of lying in bed or upon a couch day after day.

The people seemed disposed to resent the coming of the strangers at first, and declined to supply them with either food or a resting-place, till Yussuf drew out some money, and assured them that they would be paid for everything that was eaten. Then they grew more civil, and Yussuf explained to his employers that the reason for the people's churlishness was, that they were often obliged to supply food or work by some tyrannical government officer or another, and the only payment they had was in the form of blows if they complained.

The payment after they had supplied a meal of curd and milk with bread and fruit completely altered their demeanour, and upon its being intimated that a boat was required to take their visitors over to Ansina, quite a dispute arose between the owners of two as to which should have the honour and profit; but all was at length settled amicably by Yussuf, and that evening, fairly provisioned by the combined aid of the tiny village, the best of the boats hoisted its sails, and the shores of Cyprus began to look dim as the night fell, and the travellers were once more on their way.

The winds were so light and contrary that it was not until the evening of the third day that they were well in sight of the country that was to be the scene of their journeyings for many months to come; and then, as they neared Ansina, it was to see a scattered town that seemed as if of marble beyond the purple sea, while beyond the town lay to right and left a

fairy-like realm of green and gold, beyond which
again lay range upon range of amethystine mountains,
above which in turn were peaks of dazzling white,
save where the evening sun was gilding salient points
of a pure pale gold.

The run had been very pleasant in spite of the
cramped accommodation, for the little crew were a
kindly simple people, whose countenances invited trust,
and though the fare on board had been scant, yet it
was wholesome and good, as the rest the travellers had
found was grateful.

So satisfactory was this part of the trip that Mr.
Burne forgot about his back, and as he stood gazing at
the glorious panorama, indulging in an occasional pinch
of snuff, he suddenly whisked out his handkerchief and
blew a clarion blast which made the boatmen start.

"Hah!" he exclaimed suddenly; "this will do. I tell
you what it is, Preston; when I get back I shall start
a company for the reclamation of this country. It
must be taken from the Turks, and we must have a
new English colony here."

"The first Roman who saw the place must have felt
something like you do about his native land," said the
professor.

"Oh, the Romans had a colony here, had they?"

"Yes; and the Greeks before them."

"Humph!" ejaculated the old lawyer, as he let his
eyes wander from spot to spot glowing in the sinking
sun, and growing more beautiful as they advanced.
"Well, I always had, as a boy, a most decided objection
to the Greeks and Romans, and I used to wish that,
when they died out, their tongues had been buried

with them instead of being left behind to pester school-
boys; but now I am beginning to respect them, for
they must have known what they were about to settle
in such a land as this.   Lovely, eh, Lawrence?"

"Grand!" was the reply uttered in enraptured tones;
"but don't talk to me, please, I feel as if I could do
nothing else but look."

The professor smiled and joined him in drinking in
the beauty of the scene, till the little felucca sailed in
under the shelter of a large stone wall that formed
part of the ancient port. Here they found themselves
face to face with the handiwork of one of the great
nations of antiquity, this having been a city of the
Greeks, before the Romans planted their conquering
feet here, to die away leaving their broken columns,
ruined temples, and traces of their circus and aque-
ducts, among which the mingled race of Turks and
present-day Greeks had raised the shabby village,
more than town, that clustered about the port.

"Safe ashore at last," said the professor as he stepped
on to a large block of squared stone in which was
secured with lead an ancient ring. "Now, Lawrence,
our travels are to begin. How do you feel? ready for
plenty of adventure?"

"Yes, quite," was the reply.

"Then, first of all, for a comfortable resting-place.
To-morrow we will see the resident, and then make
preparations for our start."

"Humph!" ejaculated Mr. Burne; and he blew his
nose in a way never heard in Asia Minor before.

CHAPTER XVII.

PREPARATIONS FOR A JOURNEY.

AWRENCE GRANGE left England as weak and helpless in mind as he was in body; but, in the brief period that had elapsed, his mind had rapidly recovered its balance, and, leaving his body behind, had strengthened so that, eager and bright, and urged on by the glorious novelty of the things he saw, his spirit was now always setting his body tasks that it could not perform.

"I'm sure I am getting worse," he said one morning, after returning from having a delicious bathe down by the ruins of the old port. "I never felt so weak as this in England."

The professor burst into a hearty fit of laughter, in which the old lawyer joined, and then took snuff and snapped his fingers till both his companions sneezed.

"I say," cried Lawrence, "isn't it cruel of you two, laughing at a poor fellow for what he cannot help."

He looked so piteously at them that they both grew serious directly.

"Why, my dear boy," cried Mr. Preston, "can you not see that you keep on overtasking yourself? Growing worse! Now, be reasonable; you had to be carried down to the fly in London; the porters carried you to the first-class carriage in which you went down by rail, and you were carried to the steamer."

"Yes," said Lawrence sadly; "that is true, but I did not feel so weak as this."

"Get out, you young cock-goose!" cried Mr. Burne. "Why, you have been bathing, and you haven't had your breakfast yet."

"And you are mistaking fatigue for weakness," said the professor.

"Of course," cried Mr. Burne. "Why, look here. You were out nearly all day yesterday with us or with Yussuf looking at ruins, going over the place, and seeing about the horses, and now, as soon as you woke this morning, you were off with Preston here to kick and splash about in the water. Weak? what nonsense! Oh, here's Yussuf. Here, hi! you grand Turk, what do you say about this boy? He thinks he is not so well."

"The young effendi?" cried Yussuf. "Oh! I have been out this morning to see some other horses, excellencies, that are far better than any we have yet seen. They are rough, sturdy little fellows from the mountains, and you ought to buy these."

"Buy or hire?" said the professor.

"Buy, excellency. You will feed and treat them well, and at the end they will be worth as much if not more than you gave for them. Besides, if you hire horses, they will be inferior, and you will be always changing and riding fresh beasts."

"Yes, of course," said the old lawyer; "but there is no risk."

"Your excellency will pardon me, there will be more risks. We shall traverse many dangerous mountain paths, and a man should know his horse and his horse know him. They should be good friends, and take care of each other. A Turkish horse loves the hand that feeds him, the master that rides upon his back."

" I am sure you are right, Yussuf," said the professor. " We will go by your advice and buy the horses."

" Here, hold hard!" cried Mr. Burne. " Look here. Do you mean to tell me that I am expected to ride a horse along a dangerous mountain road? I mean a shelf over a precipice."

" Certainly, your excellency, the roads are very bad."

" You do not feel nervous about that, do you, Burne?" said the professor.

" Oh, dear me, no, not at all," cried the old lawyer sarcastically. "Go on. I've had a pretty good hardening already, what with knocking on the head, drowning, shipwrecking, starving, and walking off my legs."

" But, if you really object to our programme, we will try some easier route," said the professor.

" Oh, by no means, sir, by no means. I have only one thing to say. I see you have made up your mind to kill me, and I only make one proviso, and that is, that you shall take me back to England to bury me decently. I will not—I distinctly say it—I will not stay here."

" Your excellency shall come to no harm," said Yussuf, "if I can prevent it. With care and good horses there is very little risk."

" How soon shall we go to see the horses?" cried Lawrence eagerly.

" When you have been lying up for a month," replied Mr. Burne gruffly. "You are too weak, and going back too much to venture out any more."

" Till you have had a good breakfast," said the professor, laughing as he saw the lad's look of keen disappointment; and they sat down at once to a capital meal.

For they had been a week in Ansina, and were comfortably lodged in the house of a Turk whom Yussuf had recommended, and who, in a grave way, attended carefully to their wants. The luggage sent on by steamer had arrived safely, and, with the exception of the few things lost in the felucca, they were very little the worse for their mishap.

So far they had been delayed by the difficulty of obtaining horses, but now the opportunity had come for obtaining what was necessary, walking being out of the question, and the only means of traversing the rugged country, that was to be the scene of their ramblings, was by the help of a sure-footed horse.

Lawrence forgot all about his weakness as soon as breakfast was over, and started off with his companions to see the animals that were for sale.

They were at an outlying place a couple of miles away from their lodgings, and the walk in the delicious autumn air was most enjoyable. In the distance was the mysterious soft blue range of mountains that they were to penetrate for some six weeks, before the season grew too advanced, and to Lawrence it was a perfect wonderland that was to prove full of sights that would astound, adventures that would thrill; and, could he have had his way, he would have set off at once, and without all the tedious preparations that Yussuf deemed necessary.

The first mile of their way was uninteresting. Then they entered a little valley with precipitous sides,

their path running by the side of a beautiful little
stream, which they had to cross again and again; but
their progress was not rapid, for Mr. Burne always
stopped to examine the pools and talk about how fond
he had been of fishing when he was a boy.

Farther on they kept coming to little houses pleasantly situated in gardens, very much as might be seen
in the suburbs of an English town, for these were the
country houses of the wealthy Turks of the place, who
came and dwelt here in the hot times of the summer.

There was a great similarity about these places.
Houses and walls were built of fine, large, well-squared
blocks of stone and marble, with every here and there
a trace of carving visible—all showing that the Turk's
quarry was the ruined Roman city, which offered an
almost inexhaustible supply.

These little estates were either just above the river,
perched on one side, or so arranged that the stream ran
right through the grounds, rippling amongst velvety
grass lawns, overshadowed by great walnuts, with mul-
berry and plum trees in abundance.

"Hi, stop a moment," cried Mr. Burne, as they
reached one beautiful clump of trees, quite a grove,
whose leaves were waving in the soft mountain breeze.

"What have you found?" said the professor, as
Lawrence hurried up.

"That, sir, that," cried Mr. Burne. "See these trees."

"Yes," said the professor, "a magnificent clump of
planes—what a huge size!"

"Exactly," said the old lawyer. "Now, do you see
what that proves?"

"What—the presence of those trees?"

" Yes, sir," said the old lawyer dogmatically. "They show, sir, that the Turk is a much-abused man. People say that he never advances, but you see he does."

"How?" said the professor, "by being too lazy to quarry stone or marble in these mountains, where they abound, and building his house out of the edifices raised by better men?"

"No, sir; by following our example, importing from us, and planting walnut-trees and these magnificent planes all about his place. Look at these! Why, I could almost fancy myself in Gray's Inn Gardens."

" My dear Burne, are you serious?"

"Serious, sir? Never more so in my life. They are beautiful."

"Yes, they are very beautiful," said the professor drily. "But I always thought that these trees were the natives of this country, and that instead of the Turks imitating us, we had seen the beauty of these trees, and transplanted some of them when young to our own land."

"Absurd!" said the old lawyer dictatorially, and he was about to say more when Yussuf stopped at a rough kind of inclosure, where a Turk was seated upon the grass beneath a shady tree smoking thoughtfully, and apparently paying no heed to the new-comers.

"The horses are here," he said; and upon being spoken to, the Turk rose, laid aside his pipe, and bowed.

It was not a long business, for Yussuf and the owner of the horses were compatriots, but Lawrence stared at the animals in dismay when he followed his companions into the inclosure. He had pictured to himself so many lovely flowing-maned creatures of Arab descent, large-eyed, wide of nostril, and with arched necks, and tails that swept the ground. He expected to see them toss up their heads and snort, and dash off wildly, but on the contrary the dozen horses that were in the inclosure went quietly on with their grazing in the most business-like manner, and when a boy was sent to drive them up, they proved to be shaggy, heavy-headed, rather dejected-looking animals, with not an attractive point about them.

"Surely you will not buy any of these, Preston," said Mr. Burne. "I do not understand horses, but those seem to be a very shabby lot."

"They are young, effendi, healthy and strong," said Yussuf gravely. "They are accustomed to the mountains, and that is what we require. Large, handsome horses, such as you see in the desert or at Istamboul, would be useless here."

"There, I am not going to doubt your knowing best," said Mr. Burne quietly; and the bargain was made, four being selected for riding, and two that were heavier and stronger for baggage animals.

Arrangements were made for the horses to be driven before them down to Ansina, and as soon as the six purchased were driven out of the inclosure their companions trotted up, thrust their heads over a bar, and whinnied a farewell, while the others seemed to be in high glee at the change. They threw up their heads and snorted; and one that was of a cream colour, and the smallest of the lot, began to display a playfulness that upset all the rest. The way he displayed his humour was by stretching out his neck, baring his teeth, and running at and biting his companions in turn—a trick which necessitated a good deal of agility, for the other horses resented the attacks by presenting their heels to their playful companion for inspection— a proceeding of which he did not at all approve.

All went well, however, the animals were safely stowed away in the stable prepared for their use, and each was soon busy at work grinding up the barley served out for his particular benefit, oats being a luxury they were not called upon to enjoy.

CHAPTER XVIII.

MR. BURNE BLOWS HIS NOSE.

"T last!" cried Lawrence, as they set off for their first incursion. Two more days had been occupied in purchasing stores, saddlery, and other necessaries for their trip, and, as the lad said, at last they were off.

The start of the party excited no surprise in the little town. It was nothing to the people there to see four well-armed travellers set off, followed by a sturdy peasant, who had charge of the two heavily-laden pack-horses, for, in addition to the personal luggage and provisions of the travellers, with their spare ammunition, it was absolutely necessary to take a supply of barley sufficient to give the horses a good feed, or two, in case of being stranded in any spot where grain was scarce.

The heat was very great as they rode on over the plain, and Mr. Burne's pocket-handkerchief was always busy either to help him sound an alarm, to wipe the perspiration from his brow, or to whisk away the flies from himself and horse.

"It's enough to make a man wish he had a bushy tail," he said, after an exasperated dash at a little cloud of insects. " Peugh! what a number of nuisances there are in the land!"

But in a short time, enjoying the beautiful prospects spread around, they rode into a wooded valley, where the trees hung low, and, as they passed under the branches, the trouble from the virulent and hungry flies grew less.

The ascent was gradual, and after a few miles the woodland part ceased, and they found themselves upon a plain once more, but from the state of the atmosphere it was evidently far more elevated than that where the town lay. Here for miles and miles they rode through clover and wild flowers that lay as thick as the buttercups in an English meadow. But in addition to patches of golden hue there were tracts of mauve and scarlet and crimson and blue, till the eyes seemed to ache with the profusion of colour.

So far the ride had been most unadventurous. Not a house had been seen after they had quitted the out-skirts of the town, nothing but waste land, if that could be called waste where the richest of grasses and clovers with endless wild flowers abounded.

At mid-day a halt was made beneath a tremendous walnut-tree growing near a spring which trickled from the side of a hill; and now the horses were allowed to graze in the abundant clover, while the little party made their meal and rested till the heat of the day was past.

Here Yussuf pointed out their resting-place for the night—a spot that lay amid the mountains on their right, apparently not far off; but the Muslim explained that it would be a long journey, and that they must not expect to reach it before dark.

After a couple of hours the horses were loaded again, and sent on first with their driver, while the travellers followed more leisurely along the faint track, for it could hardly be called a road. The second plain was soon left behind, and their way lay among the hills, valley after valley winding in and out; and as fast as one eminence was skirted others appearing, each more elevated than the last, while the scenery grew wilder and more grand.

The little horses were behaving very well, trudging along sturdily with their riders, and every hour proving more and more the value of Yussuf's choice. There was no restiveness or skittish behaviour, save that once or twice the little cream-coloured fellow which Lawrence had selected for himself and christened Ali Baba had shown a disposition to bite one of his companions. He soon gave up, though, and walked or trotted steadily on in the file, Yussuf leading, the professor coming next, then Lawrence, and Mr. Burne last.

They stopped at various points of the rising road to study the grand patches of cedars, clumps of planes low down in the valleys, and the slopes of pines, while in the groves the thrushes sang, and the blackbirds piped as familiarly as if it was some spot in

Devonshire instead of Asia Minor. Then a diversion was made here and there to examine some spring or the edge of a ravine where a stream ran. There was plenty of time for this, as the two baggage horses had to be studied, and they were soon overtaken after one of these rides.

But at last a visit to a few stones on a hillside, which had evidently been a watch-tower in some old period of this country's history, took up so much time that the man with the baggage was a good hour's journey ahead; and as they reached the track once more Yussuf turned to ask the professor whether he thought the invalid could bear the motion if he led the way at a trot.

The professor turned to ask Lawrence, who replied that he believed he could, and then something happened.

The professor had hardly spoken and obtained his reply before Mr. Burne, who had been refreshing himself with a pinch of snuff, whisked out his handkerchief according to his custom.

They were now going along a valley which ran between too highish walls of rock, dotted here and there with trees—just the sort of place, in fact, where anyone would be disposed to shout aloud to try if there was an echo; but the idea had not occurred to either of the travellers, whose thoughts were bent upon overtaking the baggage animals with their stores, when quite unexpectedly Mr. Burne applied his handkerchief to his face and blew his nose.

It was not one of his finest blasts, there was less thunder in it, and more high-pitched horn-like music, but the effect was electrical.

There was an echo in that valley, and this echo took up the sound, repeated it, and seemed to send it on to a signalling station higher up, where it was caught and sent on again, and then again and again, each repetition growing weaker and softer than the last.

But only one of these echoes was heard by the travellers, for, as afore said, the effect was electrical.

The moment that blast was blown behind him, Ali Baba, Lawrence's cream-coloured horse, threw up his head, then lowered it, and lifted his heels, sending his rider nearly out of his saddle, uttered a peculiar squeal, and set off at a gallop.

The squeal and the noise of the hoofs acted like magic upon the other three horses, and away they went, all four as hard as they could go at full gallop, utterly regardless of the pulling and tugging that went on at their bits.

This wild stampede went on along the valley for quite a quarter of an hour before Yussuf was able to check his steed's headlong career; and it was none too soon, for the smooth track along the valley was rapidly giving way to a steep descent strewed with blocks of limestone, and to have attempted to gallop down there must have resulted in a serious fall.

As it was, Yussuf was only a few yards from a great mass of rock when his hard-mouthed steed was checked; and as the squeal of one had been sufficient to start the others, who had all their early lives been accustomed to run together in a drove, so the stopping of one had the effect of checking the rest, and they stood together shaking their ears and pawing the ground.

As soon as he could get his breath, Lawrence began to laugh, and Mr. Preston followed his lead, while the grave Muslim could not forbear a smile at Mr. Burne. This worthy's straw hat had been flying behind, hanging from his neck by a lanyard, while he stood up in his stirrups, craned his neck forward, and held his pocket-handkerchief whip fashion, though it more resembled an orange streak of light as it streamed behind; while now, as soon as the horse had stopped, he climbed out of the saddle, walked two or three steps, and then sat down and stared as if he had been startled out of his senses.

"Not hurt, I hope, Burne," said the professor kindly.

"Hurt, sir—hurt? Why, that brute must be mad. He literally flew with me, and I might as well have pulled at St. Paul's as try to stop him. Good gracious me! I'm shaken into a jelly."

"Mine was just as hard-mouthed," said the professor.

"Hard-mouthed? say iron-mouthed while you are about it. And look here, Lawrence, don't you make your pony play such tricks again."

"I did nothing, sir," expostulated Lawrence.

"Nonsense, sir! don't tell me. I saw you tickle him with your hand behind the saddle."

"But, Mr. Burne—"

"Don't interrupt and contradict, sir. I distinctly saw you do it, and then the nasty brute kicked up his heels, and squealed, and frightened the others."

"But, Mr. Burne—"

"Don't prevaricate, sir, I saw you, and when that brute squealed out you could hear the noise go echoing all down the valley."

"GRACIOUS ME! I'M SHAKEN INTO A JELLY," CRIED MR. BURNE.

In the most innocent manner—having his handkerchief out of his pocket—the old lawyer applied it to his nose and gave another blast, the result being that the horses nearly went off again; but Yussuf caught Mr. Burne's steed, and the professor and Lawrence managed to hold theirs in, but not without difficulty.

"What! were you doing it again?" cried Mr. Burne angrily.

"My dear Burne—no, no; pray, don't do that," cried the professor. "Don't you see that it was you who startled the animals off?"

"I startle them? I? What nonsense!"

"But indeed you did, when you blew your nose so loudly."

"Blew my nose so loudly! Did I blow my nose so loudly?"

"Did you? why it was you who raised that echo."

"I? Raised that echo? My dear sir, are you dreaming?"

"Dreaming? No! A ride like that upon a rough Turkish horse does not conduce to dreaming. My dear Burne, did you not know that you made that noise?"

"Noise? What, when I blew my nose, or when I took snuff?"

Lawrence could not contain himself, but burst into another tremendous fit of laughter, while, when the old lawyer looked up at him angrily, and then glanced at Yussuf, it was to see that the latter had turned his face away, and was apparently busily rearranging the bridle of his horse.

"But I say, Preston," said the old lawyer then, "do

you really mean to say that I made enough noise to frighten the horses? I thought it was Lawrence there tickling that biting beast of his."

"But I did not tickle him, Mr. Burne," protested Lawrence.

"Bless my heart, it's very strange! What do you say, Preston?—you don't answer me. It is very strange."

"Strange indeed that you do not recognize the fact that the tremendous noise you made in your pocket-handkerchief started the horses."

The old gentleman looked round; then at the horses; then in his handkerchief; and back at the horses again.

"I-er—I-er—I really cannot believe it possible, Preston; I blow my nose so softly," he said quite seriously. "Would you—there—don't think I slight your word—but—er—would you mind—I'm afraid, you see, that you are mistaken—would you mind my trying the horses?"

"By no means," said the professor smiling.

"I will then," said the old gentleman eagerly; and going up to the horses, yellow handkerchief in hand held loosely as if he were about to use it, he slowly advanced it to each animal's nose.

They neither of them winced, Lawrence's cream-colour going so far as to reach out and try to take hold of it with his lips, evidently under the impression that it was some delicate kind of Turkish dried hay.

"There," said Mr. Burne triumphantly; "you see! They are not frightened at the handkerchief."

" Walk behind," said the professor, "and blow your nose—blow gently."

The old gentleman hesitated for a moment, and then blew as was suggested, not so loudly as before, but a fairly sonorous blow.

The horses all made a plunge, and had to be held in and patted before they could be calmed down again.

"What ridiculous brutes!" exclaimed Mr. Burne contemptuously. "How absurd!"

" You are satisfied, then?" said the professor.

"I cannot help being," replied Mr. Burne. "Bless my heart! It is ridiculous."

" I am growing anxious, your excellencies," said Yussuf interrupting. "The time is getting on, and I want to overtake the baggage-horses. Will you please to mount, sir?"

" Bless me, Yussuf," cried Mr. Burne testily; "any-one would think that this was your excursion and not ours."

" Your pardon, effendi, but it will be bad if the night overtakes us and we have not found our baggage. Perhaps we may have to sleep at a khan where there is no food."

" When we have plenty with the baggage. To be sure. But must I mount that animal again? I am shaken to pieces. There, hold his head."

The old gentleman uttered a sigh, but he placed his foot in the stirrup and mounted slowly, not easily, for the horse was nervous now, and seemed as if it half suspected his rider of being the cause of that startling noise.

CHAPTER XIX.

ADVENTURES IN THE HILLS.

LL the result of coming among savages," grumbled Mr. Burne. "Anyone would think that the Turks had never learned the use of the pocket-handkerchief."

"I do not suppose many of them have arrived at your pitch of accomplishment," said the professor, laughing, as they rode on along the faint track in and out of the loveliest valleys, where nature was constantly tempting them to stop and gaze at some fresh beauty. But there was every prospect of darkness overtaking them before they reached the little mountain village where they were to rest for the night; and as the time went on the beauties of nature were forgotten in the all-powerful desire to overtake the driver with the two baggage-horses, laden with that which was extremely precious to so many hungry travellers, and at every turn their eyes were strained in front to look upon the welcome sight.

"Not so much as a tail," muttered Mr. Burne. "I say," he said aloud, "what's become of that baggage?"

Yussuf was understood to say that the man must have made haste, and that they would find him at the village.

But if that was what the Muslim had said, he was wrong. For when in the darkness, after what had become quite a dangerous finish to their journey along the edge of a shelf of rock, where, far below, the rushing and gurgling of a torrent could be heard, they reached the cluster of houses and the miserable khan, one thing was evident, and that was that the baggage had not arrived.

" What is to be done, Yussuf?" said the professor. " Must we go back and search for it?"

" We could do nothing in the dark, effendi," was the reply. " The path is safe enough in daylight; by night the risk is too great."

" But he may come yet," exclaimed Mr. Burne.

Yussuf only shook his head, and said that they must wait.

But he did not waste time, for he sought out the head-man of the village to ask for a resting-place for his employers, with a supply of the best food the village could afford, and barley for the horses.

The man surlily replied that they had not enough food for themselves, and that the barley had all gone to pay the taxes. They must go somewhere else.

It was now that the weary and hungry travellers found out the value of Yussuf.

For he came to the professor, as they sat together on their tired horses, and held out his hand.

" Give me the firman, excellency," he said. " These miserable people have been robbed and plundered by travellers who ask their hospitality, till they are suspicious of all strangers. Let me show the head-man the sultan's command before I use force."

The professor handed the document, and Yussuf walked straight to where the head-man was standing aloof, caught him by the shoulder and pushed him inside his house, where he made him read the order.

The effect was magical. The man became obsequious directly; the horses were led to a rough kind of stable; barley was found for them, a sturdy fellow removed bridles and saddles, and carried them into a good-sized very bare-looking room in the house, which he informed them was to be their chamber for the night.

Here a smoky lamp was soon lit; rugs were brought in, and before long a rough meal of bread, and eggs and fruit was set before them, followed by some coffee, which, if not particularly good, was warm and refreshing in the coolness of the mountain air.

The lamp burned low, and they were glad to extinguish it at last, and then lie down upon the rugs to sleep.

It seemed strange and weird there in the darkness of that room. Only a few hours before, they were in the heated plain; now by the gradual rise of the road they were high up where the mountain-breeze sighed among the cedars, and blew in through the unglazed window.

There was a sense of insecurity in being there amongst unfriendly strangers, and Lawrence realized the necessity for going about armed, and letting the people see that travellers carried weapons ready for use.

Twice over that day they had passed shepherds who bore over their shoulders what, at a distance, were taken for crooks, but which proved on nearer approach to be long guns, while each man had a formidable knife in his sash.

But, well armed though they were, Lawrence could

not trust himself to sleep. He was horribly weary, and ached all over with his long ride, but he could not rest. There was that open window close to the ground, and it seemed to him to offer great facilities for a bloodthirsty man to creep in and rob and murder, if he chose, before the sleepers could move in their own defence.

It was a window that looked like a square patch of transparent blackness, with a point or two of light in the far distance that he knew were stars. That was the danger, and he lay and watched it, listening to the breathing of his friends.

The door gave him no concern, for Yussuf had stretched himself across it after the fashion of a watchdog, and he too seemed to sleep.

How time went Lawrence could not tell, but he could not even doze, and the time seemed terribly long. His weariness increased, and, in addition, he began to feel feverish, and his skin itched and tingled as if every now and then an exquisitely fine needle had punctured it.

The restlessness and irritation ceased not for a moment, and he realized now that he must have caught some disease peculiar to the country. A fever, of course, but he knew enough of the laws of such complaints, from his long life of sickness, to feel that this was not a regular fever, for he perspired too freely, and his head was cool.

He tossed from side to side, but there was no rest, and when at last the window faded from his sight, and he became insensible to what was going on around him, he was still conscious of that peculiar irritation, that prickled and itched and stung and burned, till he dreamed that he was travelling through a stinging-nettle wood that led up to a square window, through which a fierce-looking Turk armed with pistols and dagger crept to come and rob him.

It was all dreadfully real, and, in the midst of his fear and agony, he could not help feeling that he was foolish to wish that the Guilford Street police-sergeant, whom he had so often seen stop by one particular lamp-post at the corner to speak to one of his men, would come now, for he had a sensation that this must be quite out of his beat.

And all the time the fierce-looking Turk was coming nearer, and at last seized him, and spoke in a low whisper.

He saw all this mentally, for his eyes were closed; but, as he opened them and gazed upwards, a broad band of pale light came through the square window, falling right on the stern face of the Turk as he bent over him just as he had fancied in his sleep.

For the moment he was about to speak. Then he calmed down and uttered a sigh as he realized the truth.

" Is that you, Yussuf?" he said.

" Yes," was the reply. "It is morning, and I thought you might like to see the sun rise from the mountain here."

" Yes, I should," said Lawrence, uttering another sigh full of relief; "but I am not well. I itch and burn —my neck, my face, my arms."

" Yes," said Yussuf sadly, as if speaking of a trouble that was inevitable.

"Is it a fever coming on?"

"Fever?" said Yussuf smiling; "oh, no! the place swarms with nasty little insects. These rugs are full."

"Ugh!" ejaculated Lawrence, jumping up and giving himself a rub and a shake. "How horrid, to be sure!"

Yussuf would not let him go far from the house, merely led him to a spot where the view was clear, and then let him gaze for a few minutes as the great orange globe rolled up and gilded the mists that lay in the hollows among the hills. Then he returned to the house and prepared the scanty breakfast, of which they partook before going off in search of the missing baggage-horses and their load.

Three hours were consumed in seeking out the spot where the man who had charge of the two animals had gone from his right path. It was very natural for him to have done so, for the road forked here, and he pursued that which seemed the most beaten way. Down here he had journeyed for hours, and when at last he had come to the conclusion that he had gone wrong, instead of turning back he had calmly accepted his fate, unloaded the animals, made himself a fire out of the abundant wood that lay around, and there he waited patiently until he was found.

It was a hindrance so soon after their starting; but Yussuf seemed to set so good an example of patience and forbearance that the professor followed it, and Mr. Burne was compelled to accept the position.

"We shall have plenty of such drawbacks," Mr. Preston said; "and we must recollect that we are not in the land of time-tables and express trains."

"We seem to be in the land of no tables at all, not even chairs," grumbled Mr. Burne; "but there, I don't complain. Go on just as you please. I'll keep all my complaints till I get back, and then put them in a big book."

A week of steady slow travelling ensued, during which time they were continually journeying in and out among the mountains, following rough tracks, or roads as they were called, whose course had been suggested by that of the streams that wandered between the hills. Often enough the way was the dried-up bed of some torrent, amidst whose boulders the patient little Turkish horses picked their way in the most sure-footed manner.

It was along such a track as this that they were going in single file one day, for some particular reason that was apparently known only to the professor and Yussuf. They seemed to be deep down in the earth, for the rift along which they travelled was not above twenty feet wide, and on the one side the rock rose up nearly three thousand feet almost perpendicularly, while, on the other, where it was not perpendicular, it appeared to overhang.

Now and then it opened out a little more. Then it contracted, and seemed as if ere long the sides of the ravine would touch; but always when it came to this, it opened out directly after.

The heat was intense, for there was not a breath of wind. The gully was perfectly dry, and wherever there was a patch of greenery, it was fifty, a hundred, perhaps a thousand feet above their heads.

"How much farther is it to the village where we

shall stop for the night?" said the old lawyer, pausing
to mop his forehead.

"There is no village that we shall stop at, effendi,"
said Yussuf quietly. "We go on a little more, and then
we shall have reached the remains that Mr. Preston
wishes to see."

"Bless my heart!" panted the old gentleman. "You
are killing that boy."

"I am quite well," said Lawrence smiling, "only
hot and thirsty. I want to see the ruins."

"Oh, go on," cried Mr. Burne. "Don't stop for
me."

Just then they were proceeding along a more open
and sunny part when the professor's horse in front
suddenly shied, swerved round, and darted back,
throwing his rider pretty heavily.

"Mind, sir! Take care!" shouted Yussuf.

"What's the good of telling a man to take care
when he is down?" cried Mr. Burne angrily; and he
tried to urge his horse forward, but it refused to stir,
while Lawrence's had behaved in precisely the same
manner, and stood shivering and snorting.

"Your gun, sir, quickly!" exclaimed Yussuf.

"What is it? Robbers?" cried Mr. Burne excitedly
as he handed the guide his double-barrelled fowling-
piece.

"No, sir; one of the evil beasts which haunt these
valleys and slopes. Is the gun loaded, sir?"

"Loaded? No, man. Do you suppose I want to
shoot somebody?"

"Quick, sir! - The charges!" whispered Yussuf;
and when, after much fumbling, Mr. Burne had forced

his hand into his cartridge-bag, Yussuf was closing the
breech of the gun, having loaded it with a couple of
cartridges handed by Lawrence, who had rapidly dismounted and drawn his sword.

It was evident that Mr. Preston was stunned by the
fall, for he lay motionless on one side of the ravine
among the stones.

" No, no, stop!" cried Yussuf as Lawrence was making his way towards the professor.

The lad involuntarily obeyed, and waited breathless to see what would follow, as Yussuf advanced cautiously, gun in hand, his dark eyes rolling from side to
side in search of the danger.

For some minutes he could see nothing.   Then, all
at once, they saw him raise the gun to his shoulder,
take a quick aim and fire, when the horses started, and
would have dashed off back, but for the fact that they
were arrested by the way being blocked by the baggage animals and Mr. Burne.

As the gun was fired its report was magnified a
hundredfold, and went rolling along in a series of peals
like thunder, while the faint blue smoke rose over
where Yussuf stood leaning forward and gazing at
some broken stones.

Then all at once he raised the gun again as if to fire,
but lowered it with a smile, and walked forward to
spurn something with his foot, and upon Lawrence
reaching him it was to find him turning over a black-
looking serpent of about six feet long, with a short
thin tail, the body of the reptile being very thick in
proportion to its length.   Upon turning it over the
Muslim pointed out that it had a peculiar reddish

throat, and he declared it to be of a very poisonous kind.

"How do you know it to be poisonous?" said Mr. Preston, who had, unseen by them, risen from where he had been thrown.

"Oh, Mr. Preston, are you much hurt?" cried Lawrence.

"I must say I am hurt," said the professor smiling. "A heavy man like me cannot fall from his horse and strike his head against the stones without suffering. But there, it is nothing serious. How do you know that is a poisonous snake, Yussuf?"

"I have been told of people being bitten by them, effendi, and some have died; but I should have said that it was dangerous as soon as I saw the horse shrink from it. Animals do not generally show such horror unless they know that there is danger."

"I don't think you are right about the horses," said the professor quietly, "for they are terrible cowards in their way; but I think you are right about the snake. Serpents that are formed like this, with the thick, sluggish-looking shape, and that peculiar short tail, are mostly venomous. Well, this one will do no more mischief, Burne."

"No. Nasty brute!" said the old lawyer, gazing down at the reptile after coaxing his horse forward. "What are you going to do, Yussuf?"

"Make sure that it will not bite any of the faithful," said the guide slowly; and drawing his knife he thrust the reptile into a convenient position, and, after cutting off its head, tossed the still writhing body to the side of the ravine.

(348)   L

This incident at an end, they all mounted again and rode on, Yussuf in the middle, and Lawrence and Mr. Preston, who declared himself better, on either hand, till, at the end of about an hour, the latter said quickly:

"Do you think you are right, Yussuf? These ravines are so much alike. Surely you must have made a mistake."

"If I am right," replied Yussuf, pointing forward, "there is a spring of clear water gushing out at the foot of that steep rock."

" And there is none, I think," said the professor, "or it would be running this way."

"If it did not run another, effendi," said Yussuf grimly. "Yes: I am right. There is the opening of the little valley down which the stream runs, and the ruined rock dwellings are just beyond."

If there had been any doubt as to their guide's knowledge it would have been set aside by the horses, for Mr. Burne suddenly uttered a warning shout, and, looking back, they saw the two baggage animals coming along at a sharp pace, which was immediately participated in by the rest of the horses, all trotting forward as fast as the nature of the ground would allow. to get to a patch of green that showed at the foot of a great rock; and upon reaching it, there, as Yussuf had said, was a copious stream, which came spouting out from a crevice in the rock, clear, cool, and delicious, for the refreshment of all.

The horses and baggage were left here in charge of the driver, and, following Yussuf, the little party were soon after at the foot of a very rugged precipice, the guide pointing upwards, and exclaiming:

"Behold, effendi, it is as I said."

For a few moments they all gazed upwards, seeing nothing but what appeared to be the rugged face of the cliff; but soon the eye began to make out a kind of order here and there, and that rugged ranges of stones had been built up on shelves of the rock, with windows and doors, but as far as could be made out these rock-dwellings had been roofless; and were more like fortifications than anything else, the professor said.

"Yes, effendi," said Yussuf gravely, "strongholds, but dwelling-places as well. People had to live in spots where they would be safe in those days. Are you going to climb up?"

"Certainly," was the reply.

"That is well, for up beyond there is a way to an old temple, and a number of caves where people must have been living."

"But where is the road up?" said Lawrence.

"Along that rough ledge," replied Yussuf. "I will go first. Would it not be better if the young effendi stayed below? The height is great, the road dangerous; and not only is it hot, but there are many serpents up among the ledges of the rock."

"What do you say, Lawrence?" said the professor.

"He is going to stop down with me," said Mr. Burne shortly.

"No, sir; I am going up," replied Lawrence. "I may never be able to see such wonders as these again."

"But, my dear boy, if you climb up here, I must go too," cried Mr. Burne.

"Come along, then, sir," cried Lawrence laughing;

" the place looks so interesting I would not miss going
up for the world."

"Humph! I know I shall be broken before I've
done," muttered Mr. Burne, taking out his handkerchief for a good blow; but glancing back in the direction where they had left the horses, he altered his
mind, as if he dreaded the consequences, and replacing
the silken square, he uttered a low sigh, and prepared
to climb.

CHAPTER XX.

THE ANCIENT DWELLINGS.

OOK here; stop a minute," said Mr. Burne;
" if we've got to climb up that break-neck
place, hadn't we better leave these guns
and things at the bottom, so as to have
our hands clear?"

" No—no—no," exclaimed Yussuf impatiently; "a
man in this country should never leave his weapons out
of his reach."

" Bah! what nonsense, sir! Anyone would think we
were at sea again, or in a country where there are no
laws."

" There are plenty of laws, Burne," said the professor, " but we are getting out of their reach."

" Highwaymen and footpads about, I suppose?" said
the old lawyer mockingly. " My dear sir, don't put
such romantic notions into the boy's head. This is not
Hounslow Heath. I suppose you will want to make

me believe next that there are bands of robbers close at hand, with a captain whose belt is stuck full of pistols—eh, Yussuf?"

"Oh, yes, sir," said their guide quietly. "I should not be surprised. There are plenty of brigands in the mountains."

"Rubbish, sir; stuff, sir; nonsense, sir!"

"It is true, sir," replied Yussuf sturdily.

"Then what do you mean, sir, if it is true, by bringing us into such a place as this?"

Yussuf stared at him wonderingly; and Lawrence burst into a hearty fit of laughter.

"Come, come, Burne," cried the professor; "if any-one is to blame, it is I. Of course, this country is in a very lawless state, but all we have to do is to preserve a bold front. Come along; we are wasting time."

Yussuf smiled and nodded, and led the way up over the crumbling stones, climbing and pointing out the easiest paths, till they were at the first ledge, and were able to inspect the first group of cliff dwellings, which proved to be strongly built roofless places, evidently of vast antiquity, and everywhere suggesting that the people who had dwelt in them had been those who lived in very troublous times, when one of the first things to think about in a home was safety, for enemies must have abounded on every side.

For about a couple of hours the professor examined, and climbed, and turned over stones, finding here and there rough fragments of pottery, while Mr. Burne settled himself down in a shady corner and had a nap.

Yussuf was indefatigable, moving fragments of rock and trying to contrive ways off the giddy slope to another group of the strange old edifices, to which in due time, and not without some risk, the professor and Lawrence climbed. But there was nothing more to reward them than they had found below, only that the wisdom of the choice of the old occupants was evident, for just as the professor had come to the conclusion that the people who made these their strongholds must have been at the mercy of the enemies who seized upon the spring down below in the ravine, they came upon proof that there was plenty of foresight exercised, and that these ancient inhabitants had arranged so as not to be forced to surrender from thirst.

It was Lawrence who made the discovery, for having climbed a little higher up the cliff face to a fresh ledge, he called to the professor to follow, and upon his reaching the spot, a great niche right in the cliff, deep and completely hidden, there were the remains of a roughly-made tank or reservoir, formed by simply building a low wall of stones and cement across the mouth, when it was evident that the water that came down from above in rainy weather would be caught and preserved for use.

It was all intensely interesting to everyone but Mr. Burne, who could not get up any enthusiasm on the subject of whom these people were, and excused himself from climbing higher on account of his back.

They descended at length, and Mr. Burne sighed with satisfaction; but Yussuf had more wonders of the past to show the travellers, pointing out a narrow path that ran diagonally up the side of the gully, and assuring the party that if they only made up their minds to ascend bravely there was no danger.

Again it was suggested that Mr. Burne should sit down and wait; but the only effect of this was to make him obstinate; and he started forward and followed Yussuf up the steep path.

It was decidedly dangerous in places where the stones had crumbled away, and a slip must have resulted in a terrible fall; but all got well over the perilous parts, and at last they climbed to a platform on the side of the huge rocky mass, where the low crumbling walls showed where a kind of temple had once stood. Here they had an opportunity of gazing down into a valley that was one mass of glorious verdure, through which dashed a torrent, whose waters flashed and glittered where the sunbeams pierced the overhanging trees, and made the scene one of the most beautiful they had seen.

There were more wonders yet, for the face of the rock was honey-combed with caverns which ran in a great distance, forming passages and chambers connected one with the other.

These had evidently been inhabited, for there were marks of tools showing how they had been enlarged, and curious well-like arrangements which suggested tanks; but Yussuf assured the travellers that these holes in the natural rock were used as stores for grain, this being the manner in which it was stored or buried to the present day.

"There," cried Mr. Burne, as they came out of the last cave, and stood once more upon the platform of rock by the ruins, and had a glorious panorama of the

defile below—"there, I've been as patient as can be
with you, but now it's my turn. What I say is, that
we must go back to camp at once, and have a rest and
a good lunch."

"Agreed," said Mr. Preston. "You have been
patient. What is it, Yussuf?" he cried suddenly, as
he saw the guide gazing intently down at something
about half a mile away, far along the winding defile.

"Travellers," said Yussuf; and in that wild, almost
uninhabited region, the appearance of fellow-creatures
excited curiosity.

They were only seen for a few minutes before the
party of mounted and unmounted men with their
baggage were seen to curve round a bold mass of rock,
and disappear into a narrow valley that turned off
almost at right angles to that by which they had come.

The descent proved more difficult than the ascent,
and Mr. Burne made several attempts to plunge down
or slide amongst the debris instead of trusting to his
feet; but these accidents were foreseen, and checked
by Yussuf, who went in front, and at the first sound of a
slip threw himself down and clung to the rock, making
himself a check or drag upon the old lawyer's progress.

They reached the bottom at last safely, but heated
and weary with the long and arduous descent.

Once on tolerably level ground in the bottom of the
defile, however, their progress was easy, and, with the
anticipation of long hearty drinks at the clear spring,
and a good meal from the store on the pack-horses'
backs, they strode on bravely in spite of the heat. The
track up to the cliff dwellings was passed; but now that
they were weary, the way seemed to be twice as far as

when they were going in the morning, and the defile looked so different upon the return journey that at last Lawrence asked with a wistful look whether they had missed the spring.

Yussuf smiled and replied that it was below, and not far distant now, and a few minutes later they turned an angle in the defile, and came in full view of the patch of verdure that marked its presence in the sterile stony gorge.

"Hah!" ejaculated Mr. Burne, "it makes one know the value of water, travelling in a land like this. Only fancy how clear and cold and refreshing it will be."

He nodded and smiled, for it was his custom after having been in any way unamiable to try and make up for it by pleasant remarks and jocularity.

"Yes," said Mr. Preston; "it does indeed. This mountain air, too, gives one an appetite—eh, Lawrence?"

"Is that curious feeling one has appetite?" said the lad. "I fancied that I was not well."

"But you feel as if you could eat?"

"Oh, yes; a great deal," cried the boy, "and I shall be glad to begin."

"Then it is hunger," said the professor laughing. "Eh, what?"

This last was in answer to some words uttered loudly by Yussuf, who had walked swiftly on, and entered the little depression where they had left the man with the horses.

"Gone, excellency, gone!" he cried excitedly, for the place was empty; the six horses and the man were not visible.

The little party stood gazing wonderingly at each other.

The water was there, gushing with great force from beneath the towering mass of rock; but their supply of food, their means of progression, the man whom they had engaged—where were they?

Yussuf stood with his hands clenched, and his brow contracted, gazing down at the ground.

Mr. Preston looked down the valley in the direction by which they had come that morning.

Mr. Burne took out his box, partook of a large pinch of snuff, and blew his nose violently.

Lawrence walked to the spring, stooped down, and began drinking, dipping up a little water at a time in the hollow of his hand.

Then there was a few moments' silence, and the professor spoke.

"It is very vexatious, just when we were so hungry, but it is plain enough. Something has startled the horses. Your Ali Baba, Lawrence, has been biting them, and they have all gone off back, and Hamed has followed to catch them. There, let's have a draught of spring water and trudge back."

"Humph! yes," said Mr. Burne hopefully. "We may meet them coming back before long."

They each drank and rose refreshed.

"Come, Yussuf," said the professor. "This way."

"No, effendi," he exclaimed sharply; "not that way, but this."

"What do you mean?" cried Mr. Preston, for the guide pointed up the ravine instead of down.

"The horses have not been frightened, but have been stolen—carried off."

"Nonsense, man!" cried Mr. Burne.

"See!" said Yussuf, pointing to the soil moistened by the stream that ran from the source, "the horses have gone along this little valley by the side of the stream — here are their hoof-marks — and come out again higher up beyond this ridge of the mountain. Yes: I know. The valleys join again there beyond where we were to-day, and I ought to have known it," he cried, stamping his foot.

"Known? Known what, man?" cried Mr. Burne angrily.

"That those men, who I said were travellers, were the robbers, who have seized our horses, and carried everything off into the hills."

CHAPTER XXI.

A SKIRMISH.

"THIS is a pretty state of affairs," cried Mr. Burne, opening and shutting his snuff-box to make it snap. "Now, what's to be done?"

'Tramp to the nearest village, I suppose, and buy more," replied the professor coolly. "We must expect reverses. This is one."

"Hang your reverses, man! I don't expect and I will not have them, if I can help it—serves us right for not watching over our baggage."

"Well, Yussuf, I suppose you are right," said the professor.

"Yes, effendi.   What is to be done?"

"What I say."

"Yes; what you say," replied the Turk frowning; "and he is so young.   We are only three."

"What are you thinking, Yussuf?"

"That it makes my blood boil, effendi, to be robbed; and I feel that we ought to follow and punish the dogs.   They are cowards, and would fly.   A robber always shrinks from the man who faces him boldly."

"And you would follow them, Yussuf?"

"If your excellency would," he said eagerly.

The grave quiet professor's face flushed, his eyes brightened, and for a few moments he felt as if his youthful days had come back, when he was one of the leaders in his college in athletics, and had more than once been in a town-and-gown row.   All this before he had settled down into the heavy serious absent-minded student.   There was now a curious tingling in his nerves, and he felt ready to agree to anything that would result in the punishment of the cowardly thieves who had left them in such a predicament; but just then his eyes fell upon Lawrence's slight delicate figure, and from that they ranged to the face of Mr. Burne, and he was the grave professor again.

"Why, Preston," said the old lawyer, "you looked as if you meant fighting."

"But I do not," he replied.   "Discretion is the better part of valour, they say."   Then, turning to Yussuf— "What is the nearest place to where we are now?"

Yussuf's face changed.   There was a look of disappointment in it for a few moments, but he turned grave and calm as usual, as he said:

"There is a village right up the valley, excellency. It is partly in the way taken by the robbers, but they will be far distant by now. They are riding and we are afoot."

"But is it far?"

"Half the distance that it would be were we to return to the place we left this morning."

"Forward, then. Come, Lawrence, you must walk as far as you can, and then I will stay with you, and we will send the others forward for help."

"I do not feel so tired now," said the lad. "I am ready."

Yussuf took the lead again and they set off, walking steadily on straight past the cliff dwellings, and the ruins by the cave, till they reached the spot in the beautifully-wooded vale where, from far above, they had seen the horsemen pass, little thinking at the time that they were bearing off their strong helps to a journey through the mountains, and all the food.

Here the beaten track curved off to the left, and the traces left by the horses were plain enough to see, for there was a little patch of marshy ground made by a little spring here, and this they had passed, Yussuf eagerly scanning them, and making out that somewhere about twelve horses had crossed here, and there were also the footprints of five or six men.

"If we go this way we may overtake the scoundrels," said the old lawyer; "but it will not do. Yussuf, I am a man of peace, and I should prove to be a very poor creature in another fight. I had quite enough to last me the rest of my life on board that boat. Here, let's rest a few hours."

"No, excellency; we must go on, even if it is slowly. This part of the valley is marshy, and there are fevers caught here. I have been along here twice, and there is a narrow track over that shoulder of the mountain that we can easily follow afoot, though we could not take horses. It is far shorter, too. Can the young effendi walk so far?"

Lawrence declared that he could, for the mountain air gave him strength. So they left the beaten track, to continue along a narrow water-course for a couple of miles, and then rapidly ascend the side of one of the vast masses of cliff, the path being literally a shelf in places not more than a foot wide, with the mountain on their left rising up like a wall, and on their right the rock sank right down to the stream, which gurgled among the masses of stone which had fallen from above, a couple of hundred feet below them and quite out of sight.

"'Pon my word, Yussuf, this is a pretty sort of a place!" panted Mr. Burne. "Hang it, man! It is dangerous."

"There is no danger, effendi, if you do not think of danger."

"But I do think of danger, sir. Why, bless my heart, sir, there isn't room for a man to turn round and comfortably blow his nose."

"There is plenty of room for the feet, effendi," replied Yussuf; "the path is level, and if you will think of the beautiful rocks, and hills, and listen to the birds singing below there, where the stream is foaming, and the bushes grow amongst the rocks, there is no danger."

"But I can't think about the beauty of all these things, Yussuf, my man, and I can only think I am going to turn giddy, and that my feet are about to slip."

"Why should you, effendi?" replied the Turk gravely. "Is it not given to man to be calm and confident, and to walk bravely on, in such places as this? He can train himself to go through what is dangerous to the timid without risk. Look at the young effendi!" he added in a whisper; "he sees no danger upon the path."

"Upon my word! Really! Bless my heart! I say, Preston, do you hear how this fellow is talking to me?"

"Yes, I hear," replied the professor. "He is quite right."

"Quite right!"

"Certainly. I have several times over felt nervous, both in our climb this morning, and since we have been up here; but I feel now as if I have mastered my timidity, and I do not mind the path half so much as I did."

"Then I've got your share and my own, and—now, just look at that boy. It is absurd."

"What is absurd?" said the professor quietly.

"Why, to see him walking on like that. Ill! Invalid! He is an impostor."

The professor smiled.

"I say, is it safe to let him go on like that?"

"So long as he feels no fear. See how confident he is!" said Mr. Preston.

Just then Lawrence stopped for the others to overtake him.

"Have you noticed what beautiful white stone this
is, Mr. Preston?" he said.

He pointed down at the path they were on, for every
here and there the rock was worn smooth and shiny
by the action of the air and water, perhaps, too, by the
footsteps of men for thousands of years, and was
almost as white as snow.

"Yes," said the professor, "I have been making
a mental note of it, and wishing I had a geologist's
hammer.   You know what it is, I suppose?"

"White stone, of course," said Mr. Burne.

"Fine white marble," said the professor.

"Nonsense, sir!   What! in quantities like this?"

"To be sure."

"But it would be worth a large fortune in London."

"Exactly, and it is worth next to nothing here,
because it could not be got down to the sea-shore,
and the carriage would be enormous."

"What a pity!" exclaimed the old lawyer.   "Dear
me!   Fine white marble!   So it is.   What a company
one might get up.   The Asia Minor Major Marble
Quarry Company—eh, Preston?"

"Yes, in hundred-pound shares that would be worth
nothing."

"Humph! I suppose not.   Well, never mind.   I'd
rather have a chicken pie and a loaf of bread now
than all the marble in the universe.   Let's get on."

Their progress was slow, for in spite of all that
Yussuf had said they had to exercise a great deal
of care, especially as the narrow track rose higher and

higher, till they were at a dizzy height above the little stream, whose source they passed just as the sun was getting low; and then their way lay between two steep cliffs; and next round a sunny slope that was dotted with huge walnut-trees, the soil being evidently deep and moist consequent upon a spring that crossed their path.

The trees were of great girth, but not lofty, and a peculiarity about them was that they were ill-grown, and gnarled and knotted in a way that made them seem as if they were diseased. For every now and then one of them displayed a huge lump or boss, such as is sometimes seen upon elms at home.

" There's another little fortune there, Burne," said the professor quietly.

" Nonsense, sir! There isn't a tree in the lot out of which you could cut a good board. Might do for gun-stocks."

" My dear Burne," said the professor, " don't you know that these large ugly bosses go to Europe to be steamed till they are soft, and then shaved off into leaves as thin almost as coarse brown paper, and then used and polished for all our handsome pianofortes?"

" No," said Mr. Burne shortly, " I didn't know it, and I didn't want to know it. I'm starving, and my back is getting bad again. Here, Yussuf, how much farther is it?"

" Two hours' journey, excellency; but as soon as we reach that gap in the rocks we come to a road that leads directly to the village, and the walking will be easier."

" Hadn't we better try and shoot a bird or an ani-

mal, and make a fire under those trees, and see if we
can find some walnuts? I must eat something. I cannot devour snuff."

The professor smiled.

" There is nothing to shoot," he said; "and as to the
walnuts, they are very nice after dinner with wine, but
for a meal—"

" Here, Lawrence, you are tired out, my boy," cried
Mr. Burne interrupting.

"Yes, I am very tired," said Lawrence, "but I
can go on."

"It is dreary work to rest without food," said
Yussuf, " but it might be better to get on to the spring
yonder, and pick out a sheltered place among the
rocks, where we could lie down and sleep for a few
hours, till the moon rises, and then continue our
journey."

"That's the plan, Yussuf; agreed *nem. con.*," cried
Mr. Burne.

"Perhaps it will be best," said Mr. Preston, and
they journeyed on for another half hour, till they
reached the gap which their guide had pointed out,
one which proved to be the embouchure of another
ravine, along the bottom of which meandered a rough
road that had probably never been repaired since the
Romans ruled the land.

" Let us go a little way in," said Yussuf; " we shall
then be sheltered from the wind. It will blow coldly
when the sun has set."

He led the way into a wild and awful-looking
chasm, for the shadows were growing deeper, and
to the weary and hungry travellers the place had a

strangely forbidding look, suggestive of hidden dangers. But for the calm and confident way in which Yussuf marched forward, the others would have hesitated to plunge into a gorge of so weird a character, until the sun had lightened its gloomy depths.

"I think this will do," said Yussuf, as they turned an angle about a couple of hundred yards from the entrance. "I will climb up here first. These rocks look cave-like and offer shelter. Hist!"

He held up his hand, for a trampling sound seemed to come from the face of the rocks a couple of hundred feet above them, and all involuntarily turned to gaze up at a spot where the shadows were blackest.

All except Yussuf, who gazed straight onward into the ravine.

It was strange. There was quite a precipice up there, and it was impossible for people to be walking. What was more strange, there was the trampling of horses' feet, and then it struck the professor that they were listening to the echoes of the sounds made by a party some distance in.

"How lucky!" said Mr. Burne. "People coming. We shall get something to eat."

"Hush, effendi!" said Yussuf sternly. "These may not be friends."

"What?" exclaimed Mr. Burne, cocking his gun.

"Yes; that is right, excellencies; look to your arms. If they are friends there is no harm done. They will respect us the more. If they are enemies, we must be prepared."

"Stop!" said Mr. Preston, glancing at Lawrence. "We must hide or run."

"There is time for neither, effendi," said Yussuf, taking out his revolver. "They will be upon us in a minute, and to run would be to draw their fire upon us."

"Run!" exclaimed Mr. Burne; "no, sir. As I'm an Englishman I won't run. If it was Napoleon Bonaparte and his army coming, and these were the Alps, I would not run now, hungry as I am, and I certainly will not go for a set of Turkish ragamuffins or Greeks."

"Then, stand firm here, excellencies, behind these stones. They are mounted; we are afoot."

The little party had hardly taken their places in the shadow cast by a rock, when a group of horse and footmen came into sight. They were about fourteen or fifteen in number apparently, some mounted, some afoot, and low down in that deep gorge the darkness was coming on so fast that it was only possible to see that they were roughly clad and carried guns.

They came on at a steady walk, talking loudly, their horses' hoofs ringing on the stony road, and quite unconscious of anyone being close beside the path they were taking till they were within some forty yards, when a man who was in front suddenly caught sight of the group behind the rocks, checked his horse, uttered a warning cry, and the next moment ample proof was given that they were either enemies or timid travellers, who took the party by the rocks for deadly foes.

For all at once the gloomy gorge was lit by the flashes of pretty well a dozen muskets, the rocks echoed the scattered volley, and magnified it fifty-fold, and then, with a yell, the company came galloping down, to rush past and reach the open slope beyond.

A SKIRMISH WITH A BAND OF MARAUDERS.

How it all happened neither Mr. Burne nor the professor could fully have explained. It must have been the effect of Yussuf's example, for, as the bullets flew harmlessly over the party's head, he replied with shot after shot from his revolver, discharging it at the attacking group. As he fired his second shot, Mr. Burne's fowling-piece went off, both barrels almost together, and the professor and Lawrence both fired as the group reached them, and after them, as it passed and went thundering by and down the slope out beyond the entrance to the gorge.

"Load again quickly," cried the professor; "they may return. There is one poor wretch down."

His command was obeyed, empty cartridges thrown out and fresh ones inserted; but the trampling of horses' hoofs was continued, and gradually grew more faint, as the little party descended from their improvised fort. They ran down, for something curious had occurred.

As the band of horsemen charged, their company seemed to divide in two, and the cause appeared to be this:

One of the mounted men was seen to fall from his saddle and hang by the stirrup, when his horse, instead of galloping on, stopped short, and five other horses that were seen to be riderless stopped, after going fifty yards, and cantered back to their companion and huddled round him.

"Why, there's Ali Baba!" cried Lawrence excitedly, as he ran down and caught his little steed by the bridle.

"And the pack-horses!" cried Mr. Burne quite as excitedly, as he followed.

"Enemies, not friends, effendi," said Yussuf quickly.

For all had seen at once now that they had recovered their lost horses, it being evident that the travellers, by taking the short cut, had got ahead of the marauding band, for such they seemed to be; and they had possibly made the task the easier by halting somewhere on the way to let their horses feed.

But there was another cause for the horses keeping together, and not following those of the strangers in their headlong flight, for, on coming up, the reason for the first one stopping was perfectly plain. Hamed, the pack-horse driver, had been made prisoner, and, poor fellow! secured by having his ankles bound together by a rope which passed beneath the horse's girths. When the charge had been made he had slipped sidewise, being unable to keep his seat, and gone down beneath his horse, with the result that the docile, well-trained animal stopped at once, and then its comrades had halted and cantered back.

"Is he much hurt, Preston?" said Mr. Burne eagerly, as the professor supported the poor fellow, while Yussuf drew out his dagger and cut the rope.

"I cannot say yet. Keep your eyes on the mouth of the gorge, and fire at once if the scoundrels show again."

"They will not show again, effendi," said Yussuf. "They are too much scared. That's better. The horses will stand. They know us now. Take hold of your bridle, Mr. Lawrence, and the others will be sure to stay."

Lawrence obeyed, and rested his piece on the horse's back, standing beside him and watching the mouth of the defile, while the others carried the injured man to the side and laid him down, the professor taking out his flask which was filled with spirit.

"Yes," said Yussuf, acquiescing. "It is not a drink for a true believer, but it is a wonderful medicine, effendi."

So it proved, for soon after a little had been poured down Hamed's throat the poor fellow opened his eyes and smiled.

"It is your excellencies!" he said in his native tongue; and upon Yussuf questioning him, he told them faintly that he was not much hurt, only a little stunned. That he was seated by the fount, with his horses grazing, when the band of armed men rode up, and one of them struck him over the head with the barrel of his musket, and when he recovered somewhat he found himself a prisoner, with his legs tied as he was found, and the horses led and driven down a narrow defile, out of which they had made their way into a forest of shady trees. Later on they had made a halt for a couple of hours, and then continued their journey, which was brought to an end, as far as he was concerned, by his falling beneath his horse.

"What is to be done now?" said the professor.

"Eat," exclaimed Mr. Burne, "even if we have to fight directly after dinner."

"The effendi is right," said Yussuf smiling. "If we go on, we may fall into a trap. If we go back a little way here till we find a suitable spot, the enemy will not dare to come and attack us in the dark. Can you walk, Hamed?"

The poor fellow tried to rise, but his ankles were

perfectly numbed, and there was nothing for it but to help him up on one of the horses, and go back farther into the gloomy ravine, which was perfectly black by the time they had found a likely place for their bivouac, where the horses would be safe as well, and this done, one of the packs was taken down from its bearer and a hearty meal made by all, Yussuf eating as he kept guard with Lawrence's gun, while Hamed was well enough to play his part feebly, as the horses rejoiced in a good feed of barley apiece.

CHAPTER XXII.

THE USE OF A STRAW HAT.

"HERE," said Mr. Burne, as he lit a cigar, and sat with his back to a stone; "if anybody in Fleet Street, or at my club, had told me I could have such an adventure as this, I should have said—"

Here he paused.

"What, Mr. Burne?" asked Lawrence after a time.

"Tarradiddle!" replied the old gentleman shortly, and he took out his handkerchief to blow his nose, but promptly suppressed the act, and said:

"No; wait till we get somewhere that is likely to be safe."

That word "safe" occurred to everybody in the silence of that dark and solemn gorge, whose sombre aspect was enough to daunt the most courageous; but

somehow that night, in spite of the riskiness of their
position, no one felt much alarmed.

There were several things which combined to make
them feel cheerful. One was the company, for the
knowledge of being there with a trusty companion on
either side was encouraging.

Then there was the calm confidence given by the
knowledge that their enemies had run from them like
a flock of sheep before a dog.

Lastly, there were the satisfactory sensations produced by the recovery of their horses and belongings,
and consequent enjoyment of a good meal.

Taken altogether, then, after proper arrangements
had been made to secure the horses, and for a watch
being kept, no scruple was felt about lying down to
sleep, everyone with his weapons ready for use in
case of an attack, which after all was not greatly
feared.

Lawrence wanted to take his turn at keeping guard,
but the professor forbade it.

"No," he said; "you have done your day's work.
Sleep and grow strong. You will help us best by
getting vigorous;" and hence it was that the lad lay
down in the solemn stillness of the vast place, gazing
up at the stars, which seemed dazzlingly bright in the
dark sky, and then it seemed to him that he closed his
eyes for a moment, and opened them again to see the
mountain slopes bathed in sunshine, while the birds
were twittering and piping, and the black desolate
gorge of the previous night was a scene of loveliness
such as he could not have imagined possible there.

"Shows the value of the sun, Lawrence," said the

professor laughing; "and what a fine thing it would be if some of our clever experimentalists could contrive to bottle and condense enough sunshine to last us all through the winters."

Just then Yussuf came up through the dewy grasses and flowers with Lawrence's gun over his shoulder.

"Well," said the professor, "what next—a good breakfast, and then start?"

"Yes, effendi," said the Turk, "but the other way."

"Other way?"

"Yes, effendi; the band of rascals are lying in ambush for us about a mile distant."

"Are you sure?"

Yussuf smiled.

"I went out at the mouth of the ravine to observe," he said; "and I could see nothing till, all at once, I saw a flash of light."

"Well?"

"Such a flash could only be reflected from a sword or gun."

"From water—a piece of glass—or crystal."

"No, excellency. There is no water up on the mountain slope. Pieces of glass are not seen there, and a crystal must be cut and polished to send forth such rays. The enemy are waiting for us in a depression, out there beyond the mouth of the plain, and we must go back the other way."

"Of course. It will be safer. But after a time they will follow us."

"I think I can stop that, effendi," said their guide smiling; and while the horses were being loaded, and everything was being got ready for a start, Yussuf

took out his knife, and selecting from among the
bushes a good straight stick, he cut and trimmed it
carefully till it was about the length of a gun.

This done, he climbed up the ridge that screened
them from the mouth of the gorge, and, selecting a
spot from whence a good view of the sloping plain
beyond could be obtained, he walked up and down
for a few minutes.

After this he beckoned to the professor and the
others to join him; and as soon as they were there
he drew their attention to a clump of bushes, as they
seemed, but which must have been trees, a couple of
miles away, though in that wondrously clear mountain
air the distance did not seem to be a quarter.

Mr. Burne was nearest to the guide, in his straw
hat, which he had retained in safety so far through
having it secured by a lanyard, but it was growing
very shabby, and was much out of shape from its
soaking in the sea.

The professor noticed that Yussuf—who was con-
spicuous in his red fez skull-cap, about which was
rolled a good deal of muslin in the form of a turban or
puggree—kept walking up and down on the edge of
the ridge, and pointing out to Mr. Burne the beauty
of the prospect, with the distant ranges of snow-topped
mountains, and the old lawyer kept on nodding his
satisfaction.

" Yes. Very fine—very fine," he said; " but I want
my breakfast."

" There!" exclaimed Mr. Preston suddenly. " I saw
it yonder."

" The flash of light, effendi?" said Yussuf quietly.

"Yes. And there again."

"I saw it then," said Lawrence quickly; and no one doubted now that their guide was right.

After staying there for about a quarter of an hour Yussuf suggested that as the horses were ready, breakfast should be hastily eaten and they should start. Consequently all went down, a hearty meal was made, Yussuf taking his walking to and from the ridge to guard against surprise, and then he approached Mr. Burne to request him to give up his straw hat.

"My straw hat!" exclaimed the old gentleman in astonishment.

"Yes, effendi," replied Yussuf. "I propose to fasten it, after wearing it for a few minutes and walking up and down, on one of the little bushes at the top of the ridge, and to stick this little pole out by its side."

"What! to look like a man on guard?" cried Lawrence eagerly.

"Yes," replied Yussuf. "It will keep the enemy where they are watching it for half the day, even if it does not keep them till evening before they find out their mistake."

"Then, stick your turban there," said Mr. Burne shortly.

"I would, effendi, if it would do as well, but it would not be so striking, nor so likely to keep them away. They might suspect it to be a trick; but they would never think that an English effendi would leave his hat in a place like that.

"And quite right, too," said the old lawyer with a snort. "No; I shall not expose my brains to the risk of sunstroke, sir. Bah! Pish! Pooh! Absurd!"

There was a shiver among the horses, and a disposition to start off again, for Mr. Burne blew another of his sonorous blasts; but the moment he whisked out his yellow silk flag, the others, as if by instinct, seized the horses' bridles and checked them in time.

"Pah! Bless my heart!" ejaculated the old gentleman, as soon as he saw what he had done. "Here, Lawrence, you will have to take all my pocket-handkerchiefs away till we get back to a civilized land."

"If the effendi would let me have his handkerchiefs I could make him a turban to keep off the sun, or if he would condescend to wear my fez it is at his service."

"Rubbish! Stuff!" cried Mr. Burne, taking off his battered straw hat, which looked as if he had slept in it on the previous night, if not before, and then sticking it on again at a fierce angle. "Do I look like a man, sir, who would wear a fez with a towel round it? Hang it all, sir, I am an Englishman."

Yussuf bowed.

"Why, he must think me mad, Lawrence."

"My dear Burne," said the professor smiling, "Yussuf is quite right. Come, you might make that concession."

"Sir, do I look like a man who would wear a fez with a jack-towel twisted round it?" cried Mr. Burne in the most irate manner.

"You certainly do not, my dear Burne," said the professor laughing; "but you do look like a man who would make any sacrifice for the benefit of his party."

"Ah! I thought as much," cried the old gentleman. "Now you come round me with carney. There, Yussuf, take it," he cried, snatching off his straw hat

and sending it skimming through the air. "Now,
then, what next? Do you want my coat and boots to
dress up your Guy Fawkes with? Don't be modest,
pray. Have even my shirt too while you are
about it."

He took five pinches of snuff in succession so close
to Ali Baba that the horse began to sneeze—or snort
would be the better term.

Yussuf smiled, and took off his fez, from which he
rapidly untwisted the muslin folds.

"Your excellency will condescend to wear my fez?"
he said.

"No, sir, I will not," cried Mr. Burne. "Certainly
not."

"But your excellency may suffer from sunstroke,"
said Yussuf. "I must insist."

"You must what?" cried Mr. Burne angrily.

"Insist, your excellency," replied Yussuf gravely.
"I am answerable for your safety. Your life, while I
am in your service, is more than mine."

"And yet, sir, you brought me here, along a break-
neck path, to fight robbers yesterday. Didn't they
shoot at me?"

"I could not prevent that, excellency," said Yussuf
smiling. "I can prevent you from being smitten by
the sun. Your handkerchief, please."

"Oh, all right!" exclaimed Mr. Burne ruefully. "I
suppose I am nobody at all here. Take it. Here are
two."

"Hah!" ejaculated Yussuf smiling with satisfaction,
and with all the oriental's love of bright colours, as
he took the two yellow silk handkerchiefs, and rolled

them loosely before arranging them in a picturesque fashion round his bright scarlet fez, and handing the head-dress back to Mr. Burne.

"Humph!" ejaculated that gentleman, putting it on with a comical expression of disgust in his countenance. "Here, you, Lawrence, if you dare to laugh at me, I'll never forgive you."

"Do, please, Mr. Burne," cried the lad, "for I must laugh: I can't help it."

So he did laugh, and the professor too, while the old lawyer gave an angry stamp.

"Look here," said the professor; "shall I wear the fez, and you can take my hat?"

"Stuff, sir! you know your head's twice as big as mine," cried Mr. Burne.

"Have mine, Mr. Burne," said Lawrence.

"Bah! do you think I've got a stupid little head like you have. No, I shall wear the fez, and I hope we shall meet some English people. It will be a warning to them not to come out into such wild spots as this."

The fact was that the old gentleman looked thoroughly picturesque, while Yussuf looked scarcely less so, as he rapidly turned the roll of muslin which he had taken from his fez into a comfortable white head-dress and put it on.

Then, taking the stick and the straw hat, he climbed up to the top of the ridge, where they saw him shoulder the stick and walk to and fro as if on guard, before rapidly arranging the hat upon the top of a little cypress-tree, and placing the stick through the branches at a slope.

So cleverly was this done, that even from where the travellers stood just below, the ruse was effective. Seen from a quarter of a mile away it must have been just like Mr. Burne on sentry.

"There," said the old lawyer with comic anger, "worse and worse. I am being set up in effigy for these barbarians to laugh at."

"No," said the professor, "we are having the laugh at them."

Yussuf came down smiling after finishing his task, and then, a final glance round having been given, and a look at the arms, they prepared to mount.

One of the baggage horses bore the grain used for their supply, and as a good feed for six horses night and morning had somewhat reduced his load, he was chosen to bear Hamed.

For the driver, in spite of the bold face he put upon the matter, was quite unfit to walk. The rough treatment he had received when his legs were tied together had completely crippled him, and in addition his head was injured by a kick from his horse when he fell.

The man was brave, though, as soon as he found that he was not to be left behind, and all being now ready, Yussuf climbed the ridge once more to see whether the enemy was approaching, and after peering just over the edge, he descended, and they went on down the defile as fast as their horses could walk.

CHAPTER XXIII.

THE PROFESSOR IS STARTLED.

T was an exciting flight, the more so from the fact that they were obliged to keep on at a foot-pace because of the baggage horses, when at any moment they knew that the enemy might appear behind in full chase. Certainly the road was bad, and it was only here and there that they could have ventured upon a trot or canter; but this did not lessen the anxiety that was felt.

A dozen times over the professor would have been glad to pause and investigate some wonderful chasm or rift, but Yussuf was inexorable. He pointed out that it would be madness to stop, for at any time the enemy might appear in sight, so Mr. Preston had to resign himself to his fate.

It was the same when, during the heat of the afternoon, they came to the ruins of a tower placed upon an angle in the defile quite a thousand feet above the rough track, so as to command a good view in every direction. From where they stood it looked ancient enough to have been erected far back in the days when the armies of Assyria or Egypt passed through these gates of the country; certainly it was not later than the Roman times.

"One might find inscriptions, perhaps, or something else to explain when it was made," said the professor. "Come, Yussuf, don't you think we might stop and ascend here?"

N

"No, effendi," replied Yussuf sternly. "Those dogs may be close upon our track, and I cannot let you run risks We are not all men."

"Yussuf is perfectly right," said Mr. Burne, who had become quite reconciled to his fez with its gaudy roll of yellow silk; in fact, two or three times over he had taken it off and held it up to examine it as it rested on his fist. "He is perfectly right," he repeated, "we do not want to fight, unless driven to extremities, and discretion is the better part of valour."

"Yes," said the professor, looking up longingly at the watch-tower, "but—"

"Now, my dear Preston, you really must not run risks for the sake of a few stones," cried the old lawyer. "Come.'

There was no help for it, so the professor sighed, and they rode slowly on, with the heat growing more and more intense, till toward sundown, when, about a hundred and fifty feet above the path, there was a cluster of ruins, evidently of quite modern date, and among them a few old fruit-trees, one of which, a plum, showed a good many purple fruit here and there.

The lawyer made a peculiar noise with his mouth as he drew rein, the others following his example.

"Now, there are some ruins that you might very well examine," he said, pointing upwards with the barrel of his gun. "Shall we dismount and climb up?"

"To see these?" said the professor quietly; and then a change came over his countenance, and he laughed softly as he turned round to look his travelling companion in the face. Which stones do you want to look at?" he said.

"Those, sir, those," cried Mr. Burne fiercely. "Can't you see?"

"No," said the professor smiling; "I do not know which you mean, whether it is the building stones or the plum stones."

"Tchah!" ejaculated the old gentleman, with his face puckering up into a comical grin. "There, come along."

Yussuf smiled too as he rode on, and at the end of a few moments he said gravely:

"The plums would not have been worth gathering, effendi. They are a bitter, sour kind."

"Grapes are too, when the fox cannot reach them —eh, Lawrence?"

No more was said, for every one was exhausted with the long slow ride. The little wind there was came from behind, and they were wandering in and out to such an extent that the soft mountain-breeze was completely shut off, and the horses were beginning to suffer terribly now from want of water to quench their burning thirst.

At last, in front, that for which they had been hoping to see appeared to be at hand, for a patch of broad green bushes at the foot of a rock told plainly that their fresh growth must be the result of abundant watering at the roots, and, pressing onward, to their delight the horses proved the correctness of their belief by breaking into a canter, and soon carrying them to where the defile ended in one of larger extent, at whose junction a spring of clear water gushed from the foot of a rock, and Lawrence cried eagerly:

"Why, this is the old place where we left Hamed!"

And so it proved to be.

Here, pursued or not, it was absolutely necessary to stop and recruit the horses, even if they had been prepared to suffer themselves; so a halt was made, one of the party took it in turn to be sentry, and the package containing provision was undone, the horses finding plenty of herbage to satisfy their wants.

Yussuf took the first watch, while Lawrence and his friends were enjoying their repast with the hunger and appetite produced by such a long fast; and then Lawrence took his place, while Yussuf seated himself upon a stone by the spring, and began eating his simple meal of hard bread and a few dates.

The night was coming on fast; and, enticed by the beauty of the shadows that were deepening in the gorge through which they had gone in pursuit of the robbers the day before, the professor walked on and on till he was nearly abreast of the rock dwellings.

They were just visible, but where he stood the gorge was in profound darkness, and he remained watching the ruins fade away as it were in the evening gloom, till, feeling that it was time to return, he was in the act of going back, when a peculiar click struck his ear, and he knew as well as if he had seen the act that a horse had struck its armed hoof against a stone.

Had he felt any doubt it was set aside by a low snort, and, feeling that one of their steeds had strayed after him, and then gone on toward the end of the gorge, he was about to hurry forward and seize it, when a second click startled him, and in an instant he realized that the enemy had evidently been duped by the sham sentry, and given up the attempt to attack them. What was more, he grasped that the enemy had started a ruse of their own, and were coming along the larger gorge, to turn back during the night by the spring, so as to take them in the rear, while they were expecting an attack in front.

The professor realized all this as he stood there in the darkness leaning upon his gun, and afraid to stir, for he knew that to do so was to betray his whereabouts to a set of men who would perhaps take his life, and even if they spared this, carry him off to hold him to ransom.

Worse still; they would then go on and surprise the party by the spring, his presence betraying their whereabouts, for there was only one spot likely in that stony wilderness for people to halt, and that was of course by the water side.

What was he to do?

It was a hard question, and the professor felt himself at his wits' end. He had stepped a dozen yards out of the track, and was standing amongst some rough stones which helped the darkness to conceal his presence, though the valley was in such a deep shadow that, as he strained eyes and ears to make out and count the enemy, he could do neither, though he knew now that they had halted just opposite to him, and he could hear them whispering evidently in consultation before they took another step in advance.

The professor stood there in the darkness with the perspiration streaming down his face as he recalled the stories he had heard of the atrocities committed by the outlaws who made their homes in the mountains of the sultan's dominions. He was tortured by a dozen different plans which suggested themselves for

his next course of action, but neither of them com-
mended itself for second consideration, while there he
was, face to face with the one great difficulty, that he
was cut off from his companions, and unable to stir
without betraying his presence and being captured or
perhaps slain.

To stir was impossible.   He hardly dared to breathe,
while his heart throbbed with so audible a beat that
he fully expected it to betray his whereabouts.

It was a perilous time, and his agony of mind was
terrible, for just then it seemed to him that he had, to
gratify his own selfishness, brought the son of his old
friend—a lad weak and wasted from a long illness—
into a peril which might have been avoided.   There
they were, perfectly unconscious of danger in this
direction; and as soon as the party had finished their
whispered consultation he felt that they would steal
cautiously on and make their attack.

What should he do—fire at them or over them, and
in the confusion make a dash for the little camp?

He dared not risk it, for it seemed a clumsy, gamb-
ling experiment, which would most probably result
in failure.

What should he do then—sacrifice himself?

Yes.   It seemed after all that his firing would not
be so clumsy an expedient, for even if it ended in his
own destruction it would warn his friends and place
them upon their guard.

He hesitated for a few moments, as he tried once
more to realize the position.   This might not, after all,
be the gang of men who had stolen their horses; but
everything pointed to the fact that it was, as he had

at first imagined — that they had been duped by Yussuf's ruse, and then made, by some way known to them, for the principal gorge, down which they had come to turn into the lesser ravine by the spring, and then in the night or early morning, take their victims in the rear, drive them out into the open country, and master them with ease.

While Mr. Preston was running over all this in his own mind he could hear the low whispering of the little body of men going on, and every now and then an impatient stamp given by one of the horses, followed by a low muttered adjuration in the Turkish dialect, bidding the animal be still.

It was only a matter of minutes, but it seemed to be hours before the band of men began to move forward cautiously through the darkness, and more than ever the professor blamed himself for not staying with his friends, but only to acknowledge the next moment that if he had done so he would not have known of the approach of the foe.

As near as he could judge the enemy had about half a mile to go, and not knowing what to do Mr. Preston began to follow them cautiously, getting as near as he could while straining his eyes to make out the figures of the mounted men as they moved slowly on.

By degrees he found out that he was left a long way behind, but while quickening his pace he was compelled to do so with the greatest caution, and to walk with outstretched hands, for, though high above his head the starlight enabled him to make out the line of the high cliff against the sky, all below in that gorge was of pitchy blackness, and he had to guide himself by stepping carefully more than by the use of his eyes.

In spite of his care he was, he found, being left more and more behind, and yet he dare not hasten for fear of coming suddenly upon the rear of the party.

But at last, quite in despair, he pressed forward, trusting to his good fortune to get near enough to note their actions without being detected, so that at last he was within a very few yards, and he kept that distance till he felt that they must be very near the spring, when, as he pressed on, keeping to the path, as he believed, he suddenly found himself about to stumble over a low block of blackish stone just beneath his feet.

He tried to save himself, but he was too late, and he blundered right upon it; but instead of knocking the skin off his shins, and falling heavily, he was stricken back, for the object he had taken for a rock felt soft, sprang up, and he found, as the man, who had been stooping to bind up his rough gear, uttered a few angry words in his own tongue, that he had come upon a laggard of the party.

It was evident that in the darkness the man imagined that he was addressing a companion, for he gripped the professor fiercely and whispered a question.

A struggle would have ensued, but just then a clear voice rang out on the night air, sounding wild and strange, and echoing from the face of the cliff as it seemed to cut the black darkness.

The man dropped the professor's arm which he had seized, sprang away into the darkness ahead, and then there was utter silence.

CHAPTER XXIV.

RECEIVING THE ENEMY.

AWRENCE kept the watch in the ravine by which they had reached the spring that day, and as he posted himself a little way up the slope, where he could shelter himself behind a block of stone and gaze for some distance along the deep rift among the rocks, he could not help feeling somewhat elated by his position.

He was stiff and sore with his long ride, but the refreshment of which he had partaken and the pleasant coolness of the evening air raised his spirits, and he smiled to himself as he felt that his strength was returning, and that he was drinking in health with every breath of the pure air around.

There was something so important, too, in his position on sentry there, with a loaded gun resting upon the rock, the gun he took such pains to polish and keep free from every spot of rust. Only a short time since he was lying back in his easy-chair in Guilford Street, waited upon incessantly by Mrs. Dunn, while now he was a traveller passing through adventures which startled him sometimes, and at others thrilled him by their strangeness and peril.

"It is like reading a book," he said to himself as he stood there watching the side of the ridge high up, with its rugged masses of stone, and a feathery cypress here and there turned to orange and gold by the setting sun.

Then he went over again the skirmish of the past night, and how the robbers had been beaten off. Next he began to wonder whether the band would stop at the end of the ravine long, and soon after, having surfeited himself with gazing at the fading light in the sky and the blackening rocks that had so lately been glistening as if of gold, he began to yawn and think that he should much like to lie down and sleep off this weariness which seemed to be coming over him like a mist.

He leaned more and more upon the stone, so as to stare down the ravine, which kept growing darker and darker, till the bushes and tall feathery cypresses began to assume suspicious forms and seem to be tall watchers or crouching men coming slowly forward to the attack.

A dozen times over he felt sure that he was right, and that he ought to fire or run back and give the alarm. But a dread of being laughed at checked him; and then he seemed to see more clearly and to make out that these were not men, but after all trees and bushes upon the slope.

This gave him more confidence for a time, as the shades of evening fell fast, and all below in the deep ravine grew black, but he was startled again by a low rushing noise that came down the valley, followed by a piteous wail which sent a chill through him, and made the hands which held the gun grow moist.

"Was it the night breeze or some bird?" he asked himself, and as he was debating with himself as to whether he might not summon Yussuf or Mr. Burne to stay with him, there came a gentle crackling noise from the side of the ravine, such as might be made by some wild beast, fresh from its lair, and in search of food.

"What could it be?" he asked himself, as in spite of his determination his nervousness increased, and he realized that strength of mind is a good deal dependent upon vigour of body, and that he was far from possessing either.

What wild beast was it likely to be? He had heard of Syrian lions, but he thought that there could not be any there now; tigers he knew enough of natural history to feel would be in India; leopards in Africa. Then what was this which approached? It must be one of two things—either a hyena or a wolf.

The former he had heard was extremely cowardly, unless it had to deal with a child or a lamb; but wolves, if hungry, were savage in the extreme, and as the noise continued, he brought the muzzle of the gun to bear, and the *click, click,* made by the locks sounded so loudly in the still evening air, that the creature, whatever it was, probably a lemur or wild-cat, took alarm, bounded off, and was heard no more.

Then the heavy sleepy sensation began to resume its sway, and though the lad remained standing, his eyes closed, and he was suddenly completely overcome with fatigue and fast asleep, when he woke with a start, for a voice just behind him said:

"Well, boy, how are you getting on?" and a faint odour of snuff, sufficient to be inhaled and to make him sneeze, roused Lawrence into thorough wakefulness.

"I was getting drowsy, Mr. Burne," said Lawrence sadly.

"Enough to make you, my lad. I've had a nap since I sat down, but I'm fresh as a daisy now. I'm to relieve you, while Yussuf or the professor is to come by and by and relieve me. I say, how do you like playing at soldiers?"

"Playing at soldiers, Mr. Burne?"

"Well, what else do you call it?—mounting guard, and fighting robbers, and all that sort of thing. I'm getting quite excited, only I don't know yet whether it's true."

"It is true enough," said Lawrence laughing.

"Oh, I don't know so much about that. It doesn't seem to be possible. Couldn't believe that such things went on in these days, when people use telephones and telegraphs and read newspapers."

"It does seem strange and unreal, sir, but then so do all these beautiful valleys and mountains."

"So they do to us, my boy. Shouldn't wonder if they are all theatrical scenery, or else we shall wake up directly both of us and say, 'Lo! it was a dream.'"

Lawrence sneezed twice heavily, for it was impossible to be in Mr. Burne's company long without suffering from the impalpable dust that pervaded all his clothes; and as the old gentleman looked on with a grim smile and clapped his young companion on the shoulder, he exclaimed:

"You are right, Lawrence, my lad, it is all real, and that proves it. I never knew anyone sneeze in a dream. There, go back. Relieve guard. I'm sentry now, and I feel as if I were outside Buckingham Palace, or the British Museum, only I ought to have a black bearskin on instead of this red fez with the yellow roll round it. How does it look, eh?"

"Splendid, sir. It quite improves you," replied Lawrence.

"Get out, you young impostor!" cried the old lawyer. "There, be off. You are getting well."

Lawrence laughed and went back to the camping-place by the spring, where Hamed was bathing his ankles in the cold water, and Yussuf was diligently attending to the horses, whose legs he hobbled so as to keep them from straying away, though they showed very little inclination for this, the clear water and the abundant clover proving too great an attraction for them to care to go far.

It was rapidly getting dark now, and hearing from Yussuf that the professor had taken his gun and strolled off along the great gorge, Lawrence was disposed to follow him, but the sensation of stiffness, the result of many hours in the saddle, made him prefer to await his return. Picking out, then, a snug spot among some stones that had fallen from above, where a clump of myrtles perfumed the soft evening air, he settled himself down, and soon sank into a comfortable drowsy state, in which he listened to the *munch munch* of the horses, and a low crooning song uttered by Hamed as he finished his task of bathing his swollen ankles, and then walked up and down more strongly, pausing every now and then to stoop and rub them well.

Soon after Yussuf came to his side, and stood looking along the gorge towards where the cliff dwellings clustered on high; but it was too dark to see them now.

"It is time the effendi was back," he said. "He will not be long now. You will keep watch while I go and speak with his excellency, Burne."

"Yes, I am well awake again, now," said Lawrence, starting up. "I wish I did not grow so sleepy."

"Why?" said Yussuf gently, as he laid his hand upon the boy's arm. "I love to see you sleep, and sleep well. It is a good sign. It means that you are growing strong and well, and will some day be a stout and active man."

"Do you think so?" said Lawrence dreamily.

"I feel sure so," replied the Turk gravely. "I am not educated like you Franks from the west, but I have lived to middle age, and noticed many things. You are growing better and stronger. I will go now and come back soon. The effendi will be here then, and we two will watch, and you shall sleep."

He strode away into the gathering darkness, passing the spring, turning round by the right, and making for the spot where the sentry were posted. Here Mr. Burne showed no inclination to go back to the little camp, but stood talking to him in his dry manner, for mutual dislike was gradually changing into a certain amount of friendliness.

Meanwhile the horses went on biting off great mouthfuls of the rich clover that grew near the stream, and munched and munched up the juicy herbage as Lawrence listened and watched the pathway to see if he could catch sight of Mr. Preston returning with his gun.

It grew darker and darker still, but the professor did not come, and Lawrence began to grow drowsy again.

He fought against it, but the desire to sleep overcame him more and more. His head sank lower, and

in an instant he was dreaming that he heard that
rustling sound again of some wild animal approaching
the group of rocks where he was stationed.

Wolf—hyena—some fierce creature that was coming
steadily on nearer and nearer, till before long it would
spring upon him, and in the nightmare-like sensation
he felt as if he were struggling to get away, while it
fascinated him and held him to his place.

One—two—three—four—there were several such
creatures drawing nearer and nearer, and he could not
cry for help, only stay motionless there in his horrible
dread.

Nearer—nearer—nearer, till he fancied he could
see them in the darkness gathering themselves up to
spring, and still he could not move—still he could not
shout to his friends for help, till all at once he seemed
to make a desperate spring, and then he was awake
and staring into the thick darkness, telling himself that
it was fancy.

No; there were sounds farther up the gorge—sounds
as of some animals coming softly down, nearer and
nearer, but not wolves or hyenas. They were horses.

There was no doubt about it—horses; and now fully
awake, the lad felt filled by a new alarm. For who
could it be but an enemy stealing along in the darkness; and in the sudden alarm, he did not pause to
argue out whether it might not be travellers like themselves, but shouted in a clear ringing voice:

"Who's that?"

There was utter stillness in the deep gorge, just
broken by the gurgling of the fount as the water
gushed from below the rock; and in his alarm, startled

as much by the deep silence as he had been by the
sounds of approaching horsemen, Lawrence shouted
again:

"Who's that?" and then, hardly knowing what he
did, he raised his gun and fired.

CHAPTER XXV.

AFTER THE SCARE.

HE sides of the gorge took up the report of Lawrence's fowling-piece, and a
volley of echoes ran rapidly along the valley; but that was no echo which rang out
directly after, for there were two bright flashes, and
a couple of shots that were magnified into terrific
sounds, as they too rolled along the deep passage between the rocks.

To Lawrence they seemed to be the answer to his
fire from the enemy, and, in the excitement of the
moment, before attempting to reload, he fired again,
the flash from his piece cutting the darkness and re-
sulting in another volley of echoes.

Then there was a hoarse shout given in a commanding voice, followed by a shrill yell, and what seemed
to be quite a large body of horsemen thundered by,
while directly after, as Lawrence was trying to reload
his piece, the darkness was cut again twice over by a
couple of clear flashes, and the rocks rang out in a

series of echoes as if a company of infantry had drawn
trigger at the word of command.

Meanwhile the beating of hoofs continued, growing
more distant minute by minute, till the sounds died away.

Then they rose again as if the band were returning,
but it was only the reflected sound from the great face
of some rock which they were approaching in their
flight; and once more the noise faded, and Lawrence,
as he stood there half petrified, heard a familiar voice
shout:

"Lawrence! Lawrence, boy, are you there?"

"Yes, yes, Mr. Preston; here."

A low murmur came out of the darkness as if the
professor had spoken some words, Lawrence never
knew what, and the next minute they were together
standing listening to the sound of footsteps, and their
guide came panting up.

"What is it?" he cried.

Mr. Preston explained, and Yussuf stood thinking
for a few moments, and hit upon the solution of the
mystery at once.

"I am not worthy of my name," he cried. "I see
it all now; they must have come round this way to
surprise us."

"And we have surprised them—so it seems," said
the professor coolly. "Our firing scared them. Will
they come back?"

"Here! anyone killed? anyone killed?" cried Mr.
Burne excitedly, as he came panting up to his friends.

"I sincerely hope not," said the professor; and he
explained anew what had occurred. "But what is to
be done now, Yussuf?"

"Excellency, I hardly know what to say. If we retreat at once it is a terrible march in the dark, and we should be much at our enemies' mercy. If we stay here we are greatly exposed, but it is better to be on guard than retreating. I learned that when fighting with my people up northward against the Russ."

"You think, then, that they will come back?"

"It is impossible to say, effendi. Perhaps not to-night, but we dare not trust them. We must be prepared."

"Let us see to the horses," cried Mr. Preston. "Hamed!"

There was no reply, but, upon Yussuf shouting the name, a response came from far up the ravine, and they found that the horses were missing.

"Oh, yes; I forgot to tell you," said Mr. Burne; "they scampered up past me, when there was all that noise down below here. One of them nearly knocked me over."

They soon found that Hamed had limped off in search of the horses which had taken fright, and but for the fact that Yussuf had hobbled their forelegs, they would have galloped away.

As it was they were soon secured, and, the party being divided into two watches, a careful guard was kept by one, while the other lay down to sleep with weapons ready to hand in case of an alarm.

CHAPTER XXVI.

YUSSUF PREACHES STICK.

HERE was no further alarm that night, for the marauders had dashed off in the full belief that they were attacked in front and rear, the four shots, multiplied by the tremendous echoes from the rocks, combining with the darkness to make them believe that their enemies were, many, and they had not stopped till they were miles away. As to making a fresh attack that was the last thing in their thoughts.

The night, then, passed peacefully away, but the amount of rest obtained was very little indeed.

After lying watching some time, Lawrence had fallen asleep, and had been awakened before daybreak by the professor, so that Hamed might have some repose; but, instead of lying down, the driver went off to his horses, and when Lawrence looked along the valley at sunrise, it was to see that Yussuf had spread his praying carpet, and was standing motionless with his hands outspread toward the east.

A hasty meal was eaten, and then a fresh start made, with Yussuf in front, and the professor and Mr. Burne, who looked like some sheik or grandee in his scarlet and yellow turban, a hundred yards behind, their guns glistening in the morning sun.

The force was not strong, for, with Yussuf as advance guard, the professor and Mr. Burne as rear,

Lawrence had to form himself into the main body, as well as the baggage guard. But as this was the whole of their available strength, the most was made of it, and they rode back along the ravine as fast as they could get the baggage horses forward, momentarily expecting attack, and in the hope of seeing some travellers or people of the country, who would, for payment, give them help; but when in the afternoon they reached the spot where the old lawyer's Panama hat, perched on the top of the cypress, still kept guard, they had not seen a soul.

Mr. Burne was for recovering his hat, but yielded to good counsel, which was in favour of hastening on to the village some few miles below in the open country, before the enemy appeared.

"Just as you like," he said. "I will not oppose you, for I do not feel at all in a fighting humour to-day."

The result was that just after sundown they rode into the little village, where about thirty men stood staring at them in a sour and evil-looking manner, not one responding to the customary salute given by Yussuf.

The latter directed himself to one of the best-dressed men, standing by the door of his house, and asked where they could get barley for the horses.

The man scowled and said that there was none to be had.

Yussuf rode on to another, who gave the same answer.

He then applied to a third, and asked where a room or rooms and refreshment could be obtained, but the man turned off without a word.

Patiently, and with the calm gentlemanly manner of a genuine Turk, he applied in all directions, but without effect.

"Have you offered to pay for everything we have, and pay well, Yussuf?" said the professor, as he sat there weary and hungry, and beginning to shiver in the cold wind that swept down from the snow-capped mountains.

"Yes, excellency, but they will not believe me."

"Show them the firman," said the professor.

This was done, but the people could not read, and when they were told of its contents they shrugged their shoulders and laughed.

It was growing dark, the cold increasing, and the travellers wearied out with their journey.

"What is to be done, Yussuf?" said Mr. Preston; "we cannot stop out here all night, and we are starving."

"They are not of the faithful," said Yussuf indignantly. "I have spoken to them as brothers, but they are dogs. Look at them, effendi. They are the friends and brethren of the thieves and cut-throats whom we met in the mountains."

"Yes, we can see that, my good friend," said Mr. Burne drily; "but as we say in our country—'Soft words butter no parsneps.'"

"No, effendi, soft words are no good here," replied Yussuf; and he took the thick oaken walking-stick which Mr. Burne carried hanging from his saddle bow.

"What are you going to do, Yussuf?" said Mr. Preston anxiously, as he glanced round at the gathering crowd of ill-looking villagers, who seemed to take great delight in the troubles of the strangers.

"Going to do, effendi?" said Yussuf in a deep voice full of suppressed anger; "going to teach these sons of Shaitan that the first duty of a faithful follower of the Prophet is hospitality to a brother who comes to him in distress."

"But, Yussuf," said Mr. Preston anxiously.

"Trust me, effendi, and I will make them remember what it is to insult three English gentlemen travelling for their pleasure. Are we dogs that they should do this thing?"

Before Mr. Preston could interfere, Yussuf gave Hamed the bridle of his horse to hold, and, making up to the man who seemed to be the head-man of the village, and who certainly had been the most insolent, he knocked off his turban, caught him by the beard, and thrashed him unmercifully with the thick stick.

Both Mr. Preston and his companion laid their hands upon their revolvers, bitterly regretting Yussuf's rashness, and fully expecting a savage attack from the little crowd of men, several of whom were armed.

But they need not have been uneasy; Yussuf knew the people with whom he had to deal, and he went on belabouring the man till he threw himself down and howled for mercy, while the crowd looked on as if interested by the spectacle more than annoyed; and when at last, with a final stroke across the shoulders, Yussuf threw the man off, the people only came a little closer and stared.

"Now," said Yussuf haughtily, and he seemed to be some magnate from Istamboul, instead of an ordinary guide, "get up and show the English lords into a good room, help unpack the baggage, and make your people prepare food."

The man rose hastily, screwing himself about and rubbing his shoulders, for he was evidently in great pain; but he seemed to get rid of a portion thereof directly by calling up three of his people, two of whom he kicked savagely for not moving more quickly, and missing the third because he did display activity enough to get out of his way.

Then obsequiously bowing to the professor and Mr. Burne, he led the way into the best house in the village, his men holding the horses, and Yussuf stopping back to see that the baggage was taken in, and the horses carefully stabled in a snug warm place, where plenty of barley was soon forthcoming.

"Why, Yussuf's stick is a regular magician's wand," said Mr. Burne, as the master of the house showed them into his clean and comfortable best room, where he bustled about, bringing them rugs and cushions, while, from the noises to be heard elsewhere, it was evident that he was giving orders, which resulted in his sending in a lad with a tray of coffee, fairly hot and good, and wonderfully comforting to the cold and weary travellers.

"Now," said Mr. Burne, "what a chance for him to poison us and finish us off."

"Have no fear of that. The man would not injure us in that way," said the professor; "but I must confess to being rather uncomfortable, for I am sure we are in a nest of hornets."

"Hark!" said Mr. Burne, "I can hear a cizzling noise which means cooking, so pray don't let's have any prophecies of evil till the supper is over. Then, perhaps, I shall be able to bear them. What do you say, Lawrence?"

"Supper first," said the latter laughing.

"Very well, then," said Mr. Preston smiling; "we will wait till after a good meal. Perhaps I shall feel more courageous then."

"What is he doing?" said Lawrence quietly, as their host kept walking in and out, for apparently no other reason than to stare at Mr. Burne's scarlet and yellow head-dress.

"I see," said Mr. Preston quietly; "he evidently thinks Mr. Burne here is some great grandee. That fez and its adornments will be a protection to us as you will see."

"Bah!" ejaculated the old lawyer; "now you are prophesying to another tune, and one is as bad as the other. Give it up; you are no prophet. Oh, how hungry I am!"

"And I," cried Lawrence.

"Well," said the professor gravely, "to be perfectly truthful, so am I. Here, mine host," he said in Arabic, "bring us some more coffee."

The man bowed low, smiled, and left the room with the empty cups, and returned directly after with them full, and after another glance at the scarlet and yellow turban, he looked at the swords and pistols and became more obsequious than ever.

CHAPTER XXVII.

CATCHING A TARTAR.

F there had been any intention on the part of their host to deal deceitfully with them, he would have had plenty of opportunity, during about a couple of hours of the night, when it was the professor's turn to keep watch, for he fell fast asleep, and was awakened by Yussuf, who shook his head at him sadly.

Morning came bright and cheery, with the birds singing, and the view from their window exquisite. Close at hand there were the mountains, rising one above another, and rich with the glorious tints of the trees and bushes that clung to their sides, and after gazing at the glorious prospect, with the clear air and dazzling sunshine, Mr Burne exclaimed:

"Bless me! What an eligible estate to lay out in building plots! Magnificent health resort! Beats Baden, Spa, Homburg, and all these places, hollow."

"And where would you get your builders and your tenants?"

"Humph! Hah! I never thought of that. But really, Preston, what a disgraceful thing it is that such a lovely country should go to ruin! Hah! here's breakfast."

For at that moment their host came in, and in a short time good bread, butter, yaourt or curd, coffee, and honey in the comb were placed before them, and somehow, after a good night's rest, the travellers did not find the owner of the house so very evil-looking.

"Oh, no, effendi, he is not a bad fellow. He bears no malice," said Yussuf, "these men are used to it. They get so terribly robbed by everyone who comes through the village that they refuse help on principle till they are obliged to give it, when they become civil."

"He is pleasant enough this morning," said Mr. Burne. "The man seems well off, too."

"Yes, effendi, he is rich for a man of his station. And now I have news for the effendi Preston."

"News? Not letters surely?" said the professor.

"No, effendi; but there are ruins close by across the valley. An old city and burying-place is yonder, this man tells me. Nobody ever goes there, because the people say that it is inhabited by djins and evil spirits, so that no one dares to go and fetch away the stones."

The professor rubbed his hands gleefully, and Mr. Burne dropped the corners of his lips as he helped himself to some more yaourt.

"How are you getting on with this stuff, Lawrence?" he said.

"I like it," was the reply.

"So do I," said Mr. Burne grimly. "It puts me in mind of being a good little boy, and going for a walk in St. James' Park with the nurse to feed the ducks, after which we used to feed ourselves at one of the lodges where they sold curds and whey. This is more like it than anything I have had since. I say, gently, young man, don't eat everything on the table."

"But I feel so hungry up here in the mountains," cried Lawrence laughing.

"Very likely, sir," said Mr. Burne with mock austerity; "but that is no reason why you should try and create a famine in the land."

"Let him eat, Burne," said the professor; "he wants bone and muscle."

"But he is eating wax," cried Mr. Burne sharply. "Let him eat chicken bone and muscle if he likes, and the flesh as well, but that would be no reason why he should eat the feathers."

"I am only too glad to see him with a good appetite," said the professor pushing the butter towards Lawrence with a smile.

"So am I. Of course. But I draw the line at wax. Confound it all, boy! be content with the honey."

"I would," said Lawrence with his mouth full; "but it is all so mixed up."

"Humph!" ejaculated Mr. Burne. "Are you going to have a look at those old stones, Preston?"

"Most decidedly."

"In spite of the djins and evil spirits?"

"Yes," replied the professor. "I suppose they will not alarm you, Yussuf?"

The guide smiled and shook his head.

"I am most alarmed about those other evil spirits, effendi," he said smiling; "such as haunt these mountains, and who steal horses, and rob men. I think the effendi will find some curious old ruins, for this seems to have been a famous place once upon a time. There is an old theatre just at the back."

"Theatre? Nonsense!" said the old lawyer with a snort.

"I meant amphitheatre, effendi—either Greek or Roman," said Yussuf politely.

"Here, I say, Yussuf," said Mr. Burne, lowering the piece of bread which he had raised half-way to his mouth; "are you an Englishman in disguise pretending to be a Turk?"

Yussuf smiled, and then turned and arrested Mr. Preston, who was about to leave his breakfast half finished and get ready to go and see the amphitheatre.

"Pray, finish first, excellency," he said. "You will not miss it now, but in a few hours' time you will be growing faint, and suffer for want of being well prepared."

"You are right," said the professor.

The breakfast was ended, and then, while the horses were being loaded, the travellers followed their host down the steep slope which formed his garden, and then by a stiff bit of pathway to where a splendid spring of water gushed right out of the rock; and the presence of this source explained a great deal, and made plain why ruins were to be found close at hand.

In fact, they came upon dressed stones directly, and it was evident that there had been a kind of temple once close to the spring, for a rough platform remained which had been cut down level to the edge of the water. The face of the rock had been levelled too, and upon it there were remains of a rough kind of inscription, while, upon examining the dressed stones which lay here and there, several, in spite of their decay, still retained the shape which showed that they had formed portions of columns.

But, search how the professor would, he could find

nothing to show what the date of the edifice had been.

Five minutes' climbing amongst broken stones brought them to a clump of trees and bushes, mingled with which were a few white-looking fragments which looked so natural that the professor's heart sank with disappointment. The stones appeared to be live stones, as geologists call it; in other words, portions of rock which had never been disturbed.

But their host pushed on through the brambles and roses, which looked as natural as if they were in an English wilderness, only that the trees that rose beyond them were strange.

"It's all labour in vain, Yussuf," said Mr. Preston in rather a disappointed tone. "You have not seen this theatre."

"No, excellency; but the man described it so exactly, that I felt he must be right; and—yes, he is."

As he spoke, he drew aside some bushes, and they found themselves gazing across heap upon heap of loose fragments of very pure white stone that was not un- like marble, and the cause of whose overthrow had most likely been the strong growth of the abundant trees, for the roots had interlaced and undermined them till they were completely forced out of place. Beyond this chaos, that lay nearly buried in greenery, rose up one above the other what seemed to Lawrence at the first glance to be the ruins of a huge flight of steps built in a semicircular form, but which he recognized at once, from pictures which he had seen, as an amphi- theatre.

There was no mistaking it. The steps, as he had

thought them to be, were the seats of stone rising tier
above tier, now broken, mouldering, and dislodged
in many places, but in others curiously perfect.

Where they, the travellers, stood must have been
occupied by the actors, far back in the past perhaps a
couple of thousand years ago; and these remains were
all that was left to tell of the greatness of the people
who once ruled in the land—great indeed, since they
left such relics as these.

Mr. Burne said "Humph!" sat down, and lit a cigar,
while their host rested upon a stone at a short distance, to admire the scarlet and yellow turban. Yussuf
followed the professor, whose eyes flashed with pleasure, while the old lawyer muttered derisively:

"Come all the way, to see a place like this! Why, I
could have taken him to the end of Holborn in a cab,
and shown him the ruins of Temple Bar all neatly num-
bered and piled up, without all these pains."

The professor did not hear his remark, for he was
too intent upon his examination of the carefully built
place, which he was ready to pronounce of Greek workmanship; but there was no one but Yussuf to hear.
For Lawrence had noted that, where the stones lay
baking in the sun, innumerable lizards were glancing
about, their gray and sometimes green armoured skins
glistening in the brilliant sunshine, and sending off
flashes every time they moved. Some were of a brown-
ish hue clouded with pale yellow; and as they darted
in and out of the crevices and holes among the stone-
work, they raised their heads on the look-out for danger, or to catch some heedless fly before darting again
beneath the levelled stones or amongst the grass and

clinging plants which were covering them here and there.

Poisonous or not poisonous? that was the question Lawrence asked himself as he crept closer and watched the actions of the nimble bright-eyed creatures, longing to capture one or two, but hesitating.

A reference to Yussuf solved the doubt.

"Oh, no; perfectly harmless as to poison," he said; "but some of the larger ones can nip pretty sharply."

"And draw blood?"

"The largest would," he said; "but you need have no fear," he added dryly; "catch all you can. I should be careful, though, for sometimes there are snakes lurking amongst the stones, and some of them are venomous. But you know the difference between a snake and a lizard?"

"Oh, yes," cried Lawrence laughing, "that's easy enough to tell."

"Not always, effendi, when they are half hidden in the grass."

Lawrence nodded, and went away to try and stalk one of the lizards. The professor was busy making measurements and taking notes, while Mr. Burne smoked on peaceably, and the Turk, who had led them here, crouched down and stared at the scarlet and yellow turban as if it fascinated him, while overhead the sun poured down its scorching beams and there was a stillness in the air that was broken by the low buzz and hum of flies, and the deep murmur of the spring below.

Lawrence crept softly along to one white stone upon which three lizards were basking; and after a moment's hesitation thrust out his hand, making sure that he had seized one by the neck, but there were three streaks upon the white stone like so many darting shadows, and there was nothing.

"Wasn't quick enough," he said to himself, and he went softly to another stone upon which there was only one, a handsome reptile, which looked as if it had been painted by nature to imitate polished tortoise-shell.

The sun flashed from its back and seemed to be hot enough to cook the little creature, which did not stir, but lay as if fast asleep.

"I shall have you easy enough," said Lawrence, as he gradually stepped up to the place and stooped and poised himself ready for the spring.

He was not hasty this time, and the reptile was perfectly unsuspicious of danger. There was no doubt about the matter—it must be asleep. He had so arranged that the sun did not cast the shadow of his arm across the stone, and drawing in his breath, he once more made a dart at the lizard, meaning if he did not catch it to sweep it away from its hole, and so make the capture more easy.

Snatch!

A brown streak that faded out as breath does from a blade of steel; and Lawrence hurt his hand upon the lichened stone.

"I'm not going to be beaten," he said to himself. "I can catch them, and I will."

He glanced at his companions, who were occupied in the amphitheatre; and, having scared away the lizards from the stones there, the lad went outside to find that there were plenty of remains about, and nearly all of them showed a lizard or two basking on the top.

He kept on trying time after time, till he grew hot and impatient, and of course, as his most careful efforts were useless, it was only natural to expect that his more careless trials would be in vain.

He was about to give the task up in despair, when all at once he caught sight of a good-sized reptile lying with its head and neck protruded from beneath a stone, and in such a position as tempted him to have one more trial.

This time it seemed to be so easy, and the reptile appeared to be one of the kind he was most eager to capture—the silvery gray, for, as they lay upon the stones, they looked as if made of oxidized metal, frosted and damascened in the most beautiful manner.

Lawrence glanced at the ground so as to be sure of his footing among the loose stones and growth, and he congratulated himself upon his foresight. For as he peered about he saw a good-sized virulent-looking serpent lying right in his way, and as if ready to strike at anybody who should pass.

Lawrence looked round for a stone wherewith to crush the creature, but he felt that if he did this he should alarm the lizard and lose it, so he drew back and picked up a few scraps, and kept on throwing first one and then another at the serpent, gently, till he roused it, and in a sluggish way it raised its head and hissed.

Then he threw another, and it again hissed menacingly, and moved itself, but all in a sluggish manner as if it were half asleep.

Another stone fell so near, though, that it made an
angry dart with its head, and then glided out of sight.

Lawrence took care not to go near where it had
disappeared, but approached the lizard on the stone
from a little to the left, which gave him a better opportunity for seizing it.

It had not moved, and he drew nearer and nearer,
to get within reach, noting the while that its body
was not in a crack from which the creature had partly
crept, but concealed by some light fine grass that he
knew would yield to his touch.

As he was about to dart his hand down and catch it
by the neck and shoulders, he saw that it was a finer
one than he had imagined, with flattish head, and very
large scales, lying loosely over one another—quite a
natural history prize, he felt.

They were moments of critical anxiety, as he softly
extended his hand, balancing himself firmly, and holding his breath, while he hesitated for a moment as to
whether he should trust to the grass giving way as he
snatched at the body, or seize the reptile by the head
and neck, and so make sure.

He had met with so many disappointments that he
determined upon the latter, and making a quick dart
down with his hand, he seized the little creature by
the neck and head, grasping it tightly, and snatching
it up, to find to his horror that he had been deceived
by the similarity of the reptile's head, and instead of
catching a lizard he had seized a little serpent about
eighteen inches long, whose head he felt moving
within his hand, while the body, which was flat and
thick for the length, wound tightly round his wrist,

and compressed it with more force than could have been expected from so small a creature.

He had uttered a shout of triumph as he caught his prize, but his voice died out upon his lips, his blood seemed to rush to his heart, and a horrible sensation of fear oppressed him, and made the cold dank perspiration ooze out upon his brow.

For he knew-as well as if he had been told that he had caught up one of the dangerous serpents of the land.

CHAPTER XXVIII.

HOW TO DEAL WITH AN ASP.

OR some minutes Lawrence Grange stood motionless as if turned to stone, and though the sun was shining down with tremendous power, he felt cold to a degree. His eyes were fixed upon the scaly creature which he held out at arm's length, and he could neither withdraw them nor move his arm, while the reptile twined and heaved and undulated in its efforts to withdraw its head from the tightly closed hand.

The boy could think little, and yet, strange as it may sound, he thought a great deal. But it was of people who had been bitten by reptiles of this kind, and who had died in a few minutes or an hour or two at most. He could not think of the best means of disembarrassing himself of the deadly creature. He could do nothing but stand with his eyes fixed upon the writhing beast.

It was an asp. He knew it was from the descriptions he had read of such creatures, and then the desire to throw it off—as far as he could, came over him, and his nerve began to return.

But only for a moment, and he shivered as he thought of the consequences of opening his hand. He saw, in imagination, the serpent clinging tightly with its body and striking him with its fangs over and over again.

But had it not already bitten him on the hand as he held that vicious head within his palm.

That he could not tell, only that he could feel the rough head of the hideous creature, and the scales pressing into his wrist. But the probability was that the creature had not bitten him, though it was heaving and straining with all its force, which, like that of all these creatures, is remarkably great for their size.

Once, as he stood there staring wildly, a peculiar swimming sensation came over him, and he felt as if he must fall; but if he did, it occurred to him that he must be at the mercy of this horrible beast, and by an effort he mastered the giddiness and stood firm.

How long he stood there he could not tell, only that the horror of being poisoned by the reptile seemed more than he could bear, especially now that life was beginning to open out with a new interest for him, and the world, instead of being embraced by the dull walls of a sick-chamber, was hourly growing more beautiful and vast.

All at once he started as it were from a dream,

LAWRENCE GRANGE'S NARROW ESCAPE FROM AN ASP.

in which before his misty eyes the hideous little serpent was assuming vast proportions, and gradually forcing open his hand by the expansion of what seemed to
be growing into a huge head. For from just behind
him there was a hoarse cry, and then a rush of feet,
and he found himself surrounded by the professor, Mr.
Burne, Yussuf, and the Turk at whose house they
stayed.

"Good heavens, Lawrence! what are you doing?"
cried the professor.

"Hush! don't speak to him," cried Yussuf in a voice
full of authority. "Let me."

As he spoke he drew his knife from his girdle.
"Lawrence effendi," he said quickly, "has it bitten you?"

The lad looked at him wildly, and his voice was a
mere whisper as he faltered:

"I do not know."

"Tell me," cried Yussuf, "have you tight hold of
it by the head?"

There was a pause, and Lawrence's eyes seemed
fixed and staring, but at last he spoke.

"Yes."

Only that word; and as the others looked on, Yussuf
caught Lawrence's right hand in his left, and compressed it more tightly on the asp's head.

"There, effendi," he said as he stood ready with his
keen bare knife in his right hand, "the serpent is
harmless now. Take hold of it by the tail, and un-
wind it from his wrist."

A momentary repugnance thrilled Mr. Preston. Then
he seized the little reptile, and proceeded to untwine
it from its constriction of Lawrence's wrist.

It seemed a little thing to do, but it was surprising how tightly it clung, and undulated, contracting itself, but all in vain, for Mr. Preston tore it off and held it out as straight as he could get the heaving body, encouraged in his efforts by Yussuf's declaration that the head was safe.

Had it not been for his strong grasp the asp would have been torn from Lawrence's failing grasp, for he was evidently growing giddy and faint, when, placing his knife as close to the neck as he could get it, Yussuf gave one bold upward cut and divided the reptile, Mr. Preston throwing down the writhing body while the head was still held tightly within Lawrence's hand.

"Do not give way, Lawrence effendi," said Yussuf in the same stern commanding voice as he had used before. "Hold up your hand—so. That is well."

He twisted the lad's clasped hand, thumb upwards, as he spoke; and those who looked on saw a few drops of blood fall from the serpent's neck as it moved feebly, the strength being now in the body that writhed among the stones.

"Let him throw it down now," cried Mr. Preston. "He may be bitten, and we must see to him."

"No," said Yussuf; "he must not open his hand yet. The head may have strength to bite even now. A few minutes, effendi, and we will see."

He watched Lawrence curiously, and with a satisfied air, for instead of growing more faint, the lad seemed to be recovering fast—so fast, indeed, that he looked up at Yussuf and exclaimed:

"Let me throw the horrid thing away."

"It did not bite you?" said Yussuf quickly.

"No, I think not. It had no time," replied Lawrence.

Yussuf said something to himself, and then, as he retained the hand within his, he exclaimed:

"Tell us how you came to seize the dangerous beast."

"I took it for a lizard," said the lad, who was nearly himself again, and then he related the whole of the circumstances.

"Hah! An easy mistake to make," said Yussuf loosening his grasp. "Now, effendi, keep tight hold and raise your hand high like this; now, quick as lightning, dash the head down upon that stone."

Lawrence obeyed, and the asp's head fell with a dull pat, moved slightly, and the jaws slowly opened, and remained gaping.

"Let me look at your hand, Lawrence," cried Mr. Preston excitedly.

"Be not alarmed, excellency," said Yussuf respectfully, his commanding authoritative manner gone. "If the young effendi had been bitten he would not look and speak like this."

"He is quite right," said Mr. Burne, who was looking very pale, and who had been watching anxiously all through this scene. "But was it a poisonous snake?"

"One of the worst we have, effendi," said Yussuf, stooping to pick up the broad flat head of the reptile, and showing all in turn that two keen little fangs were there in the front, looking exactly like a couple of points of glass.

"Yes," said the professor, "as far as I understand natural history, these are poison fangs. Bury the dangerous little thing, or crush it into the earth, Yussuf."

The guide took a stone and turned it over—a great fragment, weighing probably a hundred pounds—and then all started away, for there was an asp curled up beneath, ready to raise its head menacingly, but only to be crushed down again as Yussuf let the stone fall.

"Try another," said the professor, and a fresh fragment was raised, to be found tenantless. Beneath this the head of the poisonous reptile was thrown, the stone dropped back in its place; and, sufficient time having been spent in the old amphitheatre, they returned to the Turk's house to get their horses and ride off to see the ruins across the stream where the djins and evil spirits had their homes.

The horses were waiting when they got back, and the village seemed empty; for the people were away for the most part in their fields and gardens. Their host would have had them partake of coffee again, and a pipe, but the professor was anxious to get over to the ruins, what he had seen having whetted his appetite; so, after paying the man liberally for everything they had had, they mounted.

Quite a change had come over their unwilling host of the previous night, for as he held Mr. Preston's rein he whispered:

"Ask the great effendi with the yellow turban to forgive thy servant his treatment last night."

"What does he say, Yussuf?" asked Mr. Preston; and Yussuf, as interpreter, had to announce that if the effendis were that way again their host would be glad to entertain them, for his house was theirs and all he had whether they paid or no.

"And tell the effendis to beware," he whispered; "there are djins and evil spirits among the old mosques, and houses, and tombs; and there are evil men— robbers, who slay and steal."

"In amongst the ruins?" said Yussuf quickly.

"Everywhere," said the Turk vaguely, as he spread out his hands; and then, with their saddle-bags and packages well filled with provisions for themselves, and as much barley as could be conveniently taken, they rode out of the village and turned down a track that led them through quite a deep grove of walnut-trees to the little river that ran rushing along in the bottom of the valley. This they crossed, and the road then followed the windings of the stream for about a mile before it struck upwards; and before long they were climbing a steep slope where masses of stone and marble, that had evidently once been carefully squared or even carved, lay thick, and five minutes later the professor uttered a cry of satisfaction, for he had only to turn his horse a dozen yards or so through the bushes and trees to stand beside what looked like a huge white chest of stone.

"Hallo, what have you found?" cried Mr. Burne, rousing up, for he had been nodding upon his horse, the day being extremely hot.

"Found! A treasure," cried the professor. "Pure white marble, too."

"There, Lawrence, boy, it's in your way, not mine. I never play at marbles now. How many have you found, Preston?"

"How many?   Only this one."

"Why, it's a pump trough, and a fine one too," cried the old lawyer.

"Pump trough!" cried the professor scornfully.

"What is it then—a cistern?   I see.   Old water-works for irrigating the gardens."

"My dear sir, can you not see?   It is a huge sarco-phagus. Come here, Lawrence. Look at the sculpture and ornamentation all along this side, and at the two ends as well. The cover ought to be somewhere about."

He looked around, and, just as he had said, there was the massive cover, but broken into half a dozen pieces, and the carving and inscription, with which it had been covered, so effaced by the action of the lichens and weather that it was not possible to make anything out, only that a couple of sitting figures must at one time have been cut in high relief upon the lid.

"Probably the occupants of the tomb," said the professor thoughtfully. "Greek, I feel sure. Here, Yussuf, what does this mean?"

He caught up his gun that he had laid across the corner of the sarcophagus, and turned to face some two dozen swarthy-looking men who had come upon them unperceived and seemed to have sprung up from among the broken stones, old columns, and traces of wall that were about them on every side.

CHAPTER XXIX.

A GAME AT MARBLES.

T was a false alarm. The people who had collected about them were not brigands, and they only carried working tools, not weapons for attack.

"Means what, Yussuf?" said Mr. Burne.

"They have come to see how you dig out the buried treasure, effendi," said the guide with a suspicion of a smile.

"Treasure! what treasure?" cried the professor.

"It is of no use to argue with them, your excellency; they of course know that, in place of there being only little villages here in the far back days, there were great cities, like Istamboul and Smyrna and Trieste, all over the country."

"Quite true; there were."

"And that these cities were occupied by great wealthy nations, whose houses and palaces and temples were destroyed by enemies, and they believe that all their golden ornaments and money lie buried beneath these stones."

"What nonsense!" cried Mr. Burne impatiently. "If you dug down here you would find bones, not gold. It is an old cemetery, a place of tombs—eh, Preston?"

"Quite right," said the professor. "Tell them that we are only looking for old pieces of sculpture and inscriptions."

"I will tell them, effendi," said Yussuf smiling; and he turned to the people who were gathered round, and repeated the professor's words in their own tongue.

The result was a derisive laugh, and one of the men, a great swarthy fellow, spoke at some length.

"What does he say, Yussuf?" said Mr. Burne.

"He asks the excellency if we think they are fools and children—"

"Yes, decidedly so," replied Mr. Burne; "but hold hard, Yussuf; don't tell them so."

"If it is likely they will believe that the Franks—"

"No, no, not Franks, Yussuf," said the professor laughing; "he said 'giaours.'"

"True, effendi; he did—If they will believe that the giaours would come from a far country, and travel here merely to read a few old writings upon some stones, and examine the idols that the old people carved."

"Well, I don't wonder at it," said Mr. Burne with a sigh as he tickled his nose with a fresh pinch. "It does seem very silly. Tell them it is not they, but we: we are the fools."

"Don't tell them anything of the kind, Yussuf," said the professor. "It is not foolish to search for wisdom. Tell them the truth. We are not seeking for treasures, but to try and find something about the history of the people who built these cities."

Yussuf turned to the country people again and delivered himself of his message, after which several of the people spoke, and there was another laugh.

"Well, what do they say now?"

"They ask why you want to know all this, effendi," replied Yussuf. "It is of no use to argue with these people. They have no knowledge themselves, and they cannot understand how Frankish gentlemen can find pleasure therein. I have travelled greatly with Englishmen, and it is so everywhere. I was with an effendi down in Egypt, where he had the sand dug away from the mouth of a buried temple, and the sheik and his people who wandered near, came and drove us away, saying that the English effendi sought for silver and gold. It was the same among the hills of Birs Nimroud, where they dig out the winged lions and flying bulls with the heads of men, and the stones are covered with writing. When we went to Petra, four English effendis and your servant, we were watched by the emir and his men; and it was so in Cyprus, when the effendi I was with—an American excellency—set men to work to dig out the carved stones and idols from a temple there—not beautiful, white marble stones, but coarse and yellow and crumbling. It is always a fight here in these lands against seeking for knowledge, effendi. It is a thing they cannot understand."

"What shall we do, then?"

"What they do, effendi, half their time—nothing."

"But they will be a nuisance," cried the professor.

"Yes, effendi," said the guide, with a shrug of the shoulders "So are the flies, but we cannot drive them away. We must be content to go on just as if they were not here."

The professor saw the sense of the argument, and for the next four hours the party were busy on that hill-slope climbing amongst the stones of the ancient city—one which must have been an important place

in its day, for everywhere lay the broken fragments
of noble buildings which had been ornamented with
colonnades and cornices of elaborate workmanship.
Halls, temples, palaces, had occupied positions that
must have made the city seem magnificent, as it rose
up building upon building against the steep slope, with
the little river gurgling swiftly at the foot.

There were the remains, too, of an aqueduct, showing a few broken arches here and there, and plainly
teaching that the water to supply the place had been
mainly brought from some cold spring high up in the
mountains.

And all the time, go where they would, the travellers
were followed by the little crowd which gaped and
stared, and of which some member or another kept
drawing Yussuf aside, and offering him a handsome
present if he would confess the secret that he must
have learned—how the Frankish infidels knew where
treasure lay hid.

They seemed disappointed that the professor contented himself by merely making drawings and copy-
ing fragments of inscriptions; but at last they all
uttered a grunt of satisfaction, rubbed their hands,
gathered closely round, and seated themselves upon
the earth or upon stones.

For the professor had stopped short at the end of
what, as far as could be traced, seemed to be one end
of a small temple whose columns and walls lay scattered
as they had fallen.

Here he deliberately took a small bright trowel from
a sheath in his belt, where he carried it as if it had
been a dagger, and, stooping down, began to dig.

That was what they were waiting for. He had come at last upon the treasure spot, and though the trowel seemed to be a ridiculously small tool to work with, they felt perfectly satisfied that it was one of the wonderful engines invented by the giaours, and that it would soon clear away the stones and soil with which the treasure was covered.

"What are you doing?" said the old lawyer, as Lawrence helped the professor by dragging out pieces of stone. "Going to find anything there?"

"I cannot say," replied the professor, who was digging away energetically, and dislodging ants, a centipede or two, and a great many other insects. "This is evidently where the altar must have stood, and most likely we shall find here either a bronze figure of the deity in whose honour the temple was erected, or its fragments in marble."

"Humph! I see," cried the old lawyer, growing interested; "but I beg to remark that the evening is drawing near, and I don't think it will be prudent to make a journey here in the dark."

"No," said the professor; "it would be a pity. Mind, Lawrence, my lad; what have you there?"

"Piece of stone," said the lad, dragging out a rounded fragment.

"Piece of stone! Yes, boy, but it is a portion of a broken statue—the folds of a robe."

"Humph!" muttered the old lawyer. "Might be anything. Not going to carry it away I suppose?"

"That depends," said the professor labouring away.

"Humph!" ejaculated Mr. Burne.

"How is it that such a grand city as this should

have been so completely destroyed, Mr. Preston?" asked Lawrence.

"It is impossible to say. It may have been by the ravages of fire. More likely by war. The nation here may have been very powerful, and a more powerful nation attacked them, and, perhaps after a long siege, the soldiery utterly destroyed it, while the ravages of a couple of thousand years, perhaps of three thousand, gave the finishing touches to the destruction, and—ah, here is another piece of the same statue."

He dragged out with great difficulty another fragment of marble which had plainly enough been carved to represent drapery, and he was scraping carefully from it some adhering fragments of earth, when Mr. Burne suddenly leaped up from the block of stone upon which he had been perched, and began to shake his trousers and slap and bang his legs for a time, and then limped up and down rubbing his calf, and muttering angrily.

"What *is* the matter, Mr. Burne?" cried Lawrence.

"Matter, sir! I've been bitten by one of those horrible vipers. The brute must have crawled up my leg and—I say, Yussuf, am I a dead man?"

"Certainly not, your excellency," replied the guide gravely.

"You are laughing at me, sir. You know what I mean. I am bitten by one of those horrible vipers, am I not?"

The professor had leaped out of the little hole he had laboriously dug, and run to his companion's side in an agony of fear.

"Your excellency has been bitten by one of these,"

said the guide quietly, and he pointed to some large ants which were running all over the stones.

"Are—are you sure?" cried Mr. Burne.

"Sure, excellency? If it had been a viper you would have felt dangerous symptoms."

"Why, confound it, sir," cried Mr. Burne, rubbing his leg which he had laid bare, "that's exactly what I do feel—dangerous symptoms."

"What? What do you feel?" cried the professor excitedly.

"As if someone had bored a hole in my leg, and were squirting melted lead into all my veins—right up my leg, sir. It's maddening! It's horrible! It's worse than—worse than—there, I was going to say gout, Lawrence, but I'll say it's worse than being caned. Now, Yussuf, what do you say to that, sir, eh?"

"Ants, your excellency. They bite very sharply, and leave quite a poison in the wound."

"Quite a poison, sir!—poison's nothing to it! Here, I say, what am I to do?"

"If your excellency will allow me," said Yussuf, "I will prick the bite with the point of my knife, and then rub in a little brandy."

"Yes, do, for goodness' sake, man, before I go mad."

"Use this," said the professor, taking a little stoppered bottle from his pocket.

"What is it—more poison?" cried Mr. Burne.

"Ammonia," said the professor quietly.

"Humph!" ejaculated the patient; and he sat down on another stone, after making sure that it did not cover an insect's nest, and had not been made the roof of a viper's home.

Quite a crowd gathered round, to the old lawyer's great disgust, as he prepared himself for the operation.

"Hang the scoundrels!" he cried; "anyone would think they had never seen an old man's white leg before."

"I don't suppose they ever have, Mr. Burne," said Lawrence.

"Why, you are laughing at me, you dog! Hang it all, sir, it's too bad. Never mind, it will be your turn next; and look here, Lawrence," he cried with a malignant grin, "this is a real bite, not a sham one. I'm not pretending that I have been bitten by a snake."

"Why, Mr. Burne—"

"Well, I thought it was, but it is a real bite. Here, you, Yussuf, hold hard—what a deadly-looking implement!" he cried, as their guide bared his long keen knife. "Look here, sir, I know I'm a dog—a giaour, and that you are one of the faithful, and that it is a good deed on your part to injure me as an enemy, but, mind this, if you stick that knife thing into my leg too far, I'll—I'll—confound you, sir!—I'll bring an action against you, and ruin you, as sure as my name's Burne."

"Have no fear, effendi," said Yussuf gravely, going down on one knee, while the people crowded round.

"Cut gently, my dear fellow," said Mr. Burne; "it isn't kabobs or tough chicken, it's human leg. Hang it all! You great stupids, what are you staring at? Give a man room to breathe—*wough!* Oh, I say, Yussuf, that was a dig."

"Just enough to make it bleed, effendi. There, that

will take out some of the poison, and now I'll touch the place with some of this spirit."

" *Wough!* " ejaculated Mr. Burne again, as the wound was touched with the stopper of the bottle. " I say, that's sharp. Humph! it does not hurt quite so much now, only smarts. Thank ye, Yussuf. Why, you are quite a surgeon. Here, what are those fellows chattering about?"

" They say the Franks are a wonderful people to carry cures about in little bottles like that."

" Humph! I wish they'd kill their snakes and insects, and not waste their time staring," said the old gentleman, drawing up his stocking, after letting the ammonia dry in the sun. " Yes; I'm better now," he added, drawing down his trouser leg. " Much obliged, Yussuf. Don't you take any notice of what I say when I'm cross."

" I never do, excellency," said Yussuf smiling gravely.

" Oh, you don't—don't you?"

" No, effendi, because I know that you are a thorough gentleman at heart."

" Humph!" said Mr. Burne, as he limped to where the professor had resumed his digging. "Do you know, Lawrence, I begin to think sometimes that our calm, handsome grave Turkish friend there, is the better gentleman of the two."

CHAPTER XXX.

A TERROR OF THE COUNTRY.

T was now evening, but instead of the air becoming cooler with the wind that blew from the mountains, a peculiar hot breath seemed to be exhaled from the earth. The stones which had been baking in the sun all day gave out the heat they had taken in, and a curious sombre stillness was over everything.

"Are we going to have a storm, Yussuf?" said Mr. Burne, as he looked round at the lurid brassy aspect of the heavens, and the wild reflections upon the mountains.

"No, excellency, I think not; and the people here seem to think the same."

"Why? They don't say anything."

"No, excellency, but if they felt a storm coming they would have long ago hurried back to their houses instead of sitting here so contentedly waiting to see the effendi dig out his treasure."

For the people had not budged an inch, but patiently watched every movement made by the travellers, crouching as it were, ready to spring forward, and see the first great find.

But the professor made no great discovery. He was evidently right about the building having been a temple, and it seemed as if an altar must have stood there, bearing a figure of which he picked up several pieces beautifully sculptured, but nothing that could be restored by piecing together; and when, wearied out, he turned to examine some other parts of the old temple, the most interesting thing that he found was a piece of column, nearly buried, and remarkable for containing two of the rounds or drums secured together by means of molten lead poured through suitable holes cut in the stones.

"There," he said at last, "I have been so deeply interested in what I have seen here, that I owe you plenty of apologies, Burne, and you too, Lawrence."

"Humph!" grunted the old lawyer, "you owe me nothing. I would as soon stop here and look about at the mountains, as go on somewhere else. My word, though, what a shame it seems that these pigs of people should have such a glorious country to live in, while we have nothing better than poor old England, with its fogs and cold east winds."

"But this peace is not perfect," said the professor. "And now, look here; we had better go back to our last night's lodgings. We can get a good meal there and rest."

"The very thing I was going to propose," said Mr. Burne quickly. "Depend upon it that man will give us a pilaf for supper."

"And without Yussuf's stick," said the professor smiling. "But come along. Let's look at the horses."

The horses were in good plight, for Yussuf and Hamed had watered them, and they had made a good meal off the grass and shoots which grew amongst the ruins.

They were now busily finishing a few handfuls of barley which had been poured for them in an old ruined trough, close to some half dozen broken pillars and a piece of stone wall that had been beautifully built; and, as soon as the patient beasts had finished, they were bridled and led out to where the professor and his friends were standing looking wonderingly round at the peculiar glare over the landscape.

"Just look at those people," cried Lawrence suddenly; and the scene below them caught their eye. For, no sooner had the professor and his companions left the coast clear than these people made a rush for the hole, which they seemed to have looked upon as a veritable gold mine, and in and about this they were digging and tearing out the earth, quarrelling, pushing and fighting one with the other for the best places.

"How absurd!" exclaimed the professor. "I did not think of that. I ought to have paid them, and made them with their tools do all the work, while I looked on and examined all they turned up."

"It would have been useless, effendi," said Yussuf. "Unless you had brought an order to the pasha of the district, and these people had been forced to work, they would not have stirred. Ah!"

Yussuf uttered a peculiar cry, and the men who were digging below them gave vent to a shrill howl, and leaped out of the pit they were digging to run shrieking back towards the village on the other slope.

For all at once it seemed to Lawrence that he was back on shipboard, with the vessel rising beneath his feet and the first symptoms of sea-sickness coming on.

Then close at hand, where the horses had so short a time before been feeding, the piece of well-built wall toppled over, and three of the broken columns fell with a crash, while a huge cloud of dust rose from the earth.

The horses snorted and trembled, and again there came that sensation of the earth heaving up, just as if it were being made to undulate like the waves at sea.

Lawrence threw himself down, while Yussuf clung to the horses' bridles, as if to guard against a stampede, and the driver stood calmly in the attitude of prayer.

Then again and again the whole of the mountain side shook and undulated, waving up and down till the sensation of sickness became intolerable, and all the while there was the dull roar of falling stones above, below, away to the left and right. Now some huge mass seemed to drop on to the earth with a dull thud, another fell upon other stones, and seemed to be broken to atoms, and again and again others seemed to slip from their foundations, and go rolling down like an avalanche, and once more all was still.

"Is it an earthquake?" said Lawrence at last in a low awe-stricken tone.

"Seems like a dozen earthquakes," said the old lawyer. "My goodness me! What a place for a town!"

And as they all stood there trembling and expecting the next shock, not knowing but the earth might open a vast cavity into which the whole mountain would plunge, a huge cloud of dust arose, shutting out everything that was half a dozen yards away, and the heated air grew more and more suffocating.

It was plain enough to understand now why it was that in the course of time this beautiful city should have been destroyed. The first disaster might have been caused by war, but it was evident that this was a region where earth disturbance was a frequent occurrence, and as time rolled by, every shock would tear down more and more of the place.

Very little was said, for though no great shock came now, there were every few minutes little vibrations beneath their feet, as if the earth was trembling from the effect of the violent efforts it had made.

Now and then they held their breath as a stronger agitation came, and once this ended with what seemed to be a throb or a sound as if the earth had parted and then closed up again.

Then came a lapse, during which the travellers sat in the midst of the thick mist of dust waiting, waiting for the next great throb, feeling that perhaps these were only the preliminaries to some awful catastrophe.

No one spoke, and the silence was absolutely profound. They were surrounded by groves where the birds as a rule piped and sang loudly; but everything was hushed as if the thick dust-cloud had shut in all sound.

And what a cloud of dust! The dust of a buried city, of a people who had lived when the earth was a couple of thousand years or more younger, when western Europe was the home of barbarians. The dust of buildings that had been erected by the most civilized peoples then dwelling in the world, and this now rising in the thick dense cloud which seemed as if it would never pass.

An hour must have gone by, and they were conscious as they stood there in a group that the mist looked blacker, and by this they felt that the night must be coming on. For some time now there had not been the slightest quiver of the ground, and in place of the horses standing with their legs spread wide and heads low, staring wildly, and snorting with dread, they had gathered themselves together again, and were beginning to crop the herbage here and there, but blowing over it and letting it fall from their lips again as if in disgust.

And no wonder, for every blade and leaf was covered with a fine impalpable powder, while, as the perspiration dried upon the exposed parts of the travellers, their skins seemed to be stiff and caked with the dust.

"I think the earthquake is over, excellencies," said Yussuf calmly. "I could not be sure, but the look of the sky this evening was strange."

"I had read of earthquakes out here," said the professor, who was gaining confidence now; "but you do not often have such shocks as these?"

"Oh, yes, effendi; it is not an unusual thing. Much more terrible than this; whole towns are sometimes swallowed up. Hundreds of lives are lost, and hundreds left homeless."

"Then you call this a slight earthquake?" said Mr. Burne.

"Certainly, excellency, here," was the reply. "It may have been very terrible elsewhere. Terrible to us if we had been standing beside those stones which fell. It would have been awful enough if all these ruins had been, as they once were, grandly built houses and temples."

"And I was grumbling about poor dear old sooty, foggy England," said Mr. Burne. "Dear, dear, dear, what foolish things one says!"

"Is not the dust settling down?" said the professor just then.

"A little, your excellency; but it is so fine that unless we have a breeze it may be hours before it is gone."

"Then what do you propose to do?" asked Mr Burne.

"What can I do, excellency, but try to keep you out of danger?"

"Yes, but how?"

"We must stay here."

"Stay here? when that village is so near at hand?"

Yussuf paused for a few minutes and then said slowly, as if the question had just been asked:

"How do we know that the village is near at hand?"

"Ah!" ejaculated the professor, startled by the man's tone.

"It was not more than two of your English miles from here, excellency, when we left the place this morning, but with such a shock there may be only ruins from which the people who were spared have fled."

"How horrible!" exclaimed Lawrence.

"Let us hope that I am wrong, effendi," said Yussuf hastily. "I only speak."

"But we cannot stay here for the night," said Mr. Burne impatiently.

"Excellency, we must stay here," said the Turk firmly. "I am your guide, and where I know the land I will lead you. I knew this country this morning, but how can I know it now? Great chasms may lie

between us and the village—deep rifts, into which in
the dust and darkness we may walk. You know
what vast gorges and valleys lie between the hills."

"Yes," replied Mr. Preston.

"Some of these have been worn down by the torrents and streams from the mountains, others have
been made in a moment by such shocks as these. I
would gladly say, 'Come on; I will lead you back
to the headman's house;' but, excellencies, I do not
dare."

"He is quite right, Burne," said the professor
gravely.

"Oh, yes, confound him: he always is right," cried
Mr. Burne. "I wish sometimes he were not. Fancy
camping out here for the night in this horrible dust
and with the air growing cold. It will be icy here by
and by."

"Yes, excellency, it will be cold. We are high up,
and the snow mountains are not far away."

"We must make the best of it, Lawrence, my boy,"
said the professor cheerily. "Then I suppose the next
thing is to select a camp. But, Yussuf, this is rather
risky. What about the asps?"

"And the ants," cried Mr. Burne with a groan. "I
can't sleep with such bed-fellows as these."

"And the djins and evil spirits," cried Lawrence.

"Ah, I don't think they will hurt us much, my
boy," said the professor.

"And there is one comfort," added Mr. Burne; "we
have left the cemetery behind. I do protest against
camping there."

"A cemetery of two thousand years ago," said the.

professor quietly. "Ah, Burne, we need not make
that an objection. But come, what is to be done?"

Yussuf answered the question by calling Hamed to
come and help unpack the horses, and all then set to
work to prepare to pass the night in the midst of the
ruins, and without much prospect of a fire being made.

CHAPTER XXXI.

ALI BABA'S FEAT.

THE night came on colder and colder, and
though Yussuf and Hamed worked hard
at cutting bushes and branches of trees,
the green wood covered with leaves obsti-
nately refused to burn, and the result was a thick
smoke, which hung about and spread amongst the
dust, making the position of the travellers worse than
before. Yussuf searched as far as he could, but he
could find no pines, neither were there any bushes
of the laurel family, or the result would have been
different.

All this while they were suffering from a nervous
trepidation that made even a heavy footfall startling,
every one being in expectation of a renewal of the
earthquake shocks.

Rugs and overcoats were taken from the baggage,
and, giving up the fire as a bad job, the little party
were huddled together for the sake of warmth, when

all at once a breeze sprang up, and in less than half an hour the mist of dust had been swept away, and the dark sky was overhead studded with countless stars.

It was even colder than before, the wind that came down from the mountains being extremely searching, and it seemed a wonder that there could be so much difference between day and night. But in spite of the cold the little party felt cheered and relieved by the disappearance of the thick mist of dust. The bright sky above them seemed to be a sign of the danger, having passed away, and suggestive of the morning breaking bright and clear to give them hope and the power of seeing any dangers that were near.

But they were not to wait till morning, for soon after the clearing away of the mist, shouts were heard in the distance, to which they responded, and the communication was kept up till a party of men appeared, who proved to be no belated set of wanderers like themselves, but about twenty of the village people under the command of the headman, come in search of them, and all ready to utter a wild cheer when they were found.

The leader explained to Yussuf that the earthquake shocks had all been on this side of the little river, the village having completely escaped. About a couple of hours after the shocks the party of people who had been digging for treasure returned to the village, and upon the headman learning that the travellers had been left up there he had organized a party to come in search.

There was no mistaking the cordiality of the headman or his joy at having found them, and after help-

ing to repack the horses he led the way back con-
fidently enough, and in the walk explained that the
mischief done was very slight. No gaps had opened,
as far as he knew, but at all events the road from the
old ruins to the village was safe.

"Your cudgel seems to have been a regular genii's
wand, Yussuf," said Mr. Burne softly. "You would not
find it have so good an effect upon Englishmen."

"It and your payments, effendi, have taught the
man that we are people of importance, and not to be
trifled with," replied Yussuf smiling; and Mr. Burne
nodded and took snuff.

In an hour they were safely back at the headman's
house, where hot coffee and then a good meal prepared
all for their night's rest amidst the warm rugs which
were spread for them; and feeling that no watch was
necessary here, all were soon in a deep sleep, Lawrence
being too tired even to think of the danger to which
they had been exposed.

Directly after breakfast next morning the headman
came to them with a very serious look upon his countenance.

The people of the village were angry, he said to
Yussuf, and were uttering threats against the strangers,
for it was due to them that the earthquake had taken
place. Every one knew that the old ruins were the
homes of djins and evil spirits. The strangers had
been interfering with those ruins, and the djins and
evil spirits had resented it.

"But," said Yussuf, "your people did more than
their excellencies."

"Yes, perhaps so," said the headman; "but they are

fools and pigs. Let the English effendis go, and not touch the ruins again."

Yussuf explained, and the professor made a gesture full of annoyance.

"Ask him, Yussuf, if he believes this nonsense."

"Not when I am with you, excellencies," he said smiling; "but when I am with my people, I do. If I did not think as they do I could not live with them. I am headman, but if they turn against me they are the masters, and I am obliged to do as they wish."

There was nothing for it but to go, and they left the village with all its interesting surroundings as soon as the horses were packed, the people uttering more than one menacing growl till they were out of hearing.

"So vexatious!" exclaimed the professor. "I feel as if we have done wrong in giving up. The firman ought to have been sufficient. We shall never find such a place again—so rich in antiquities. I have a good mind to turn back."

"No, no, effendi," said Yussuf, "it would only mean trouble. I can take you to fifty places as full of old remains. Trust to me and I will show you the way."

They journeyed on for days, finding good, bad, and indifferent lodgings. Sometimes they were received by the people with civility, at others with suspicion, for Yussuf was taking them farther and farther into the mountains, where the peasants were ignorant and superstitious to a degree; but, save where they crossed some plain, they were everywhere impressed by the grandeur of the country, and the utter ruin and neglect which prevailed. Roads, cities, land, all seemed to have been allowed to go to decay; and, to make the journey the longer and more arduous, over and over again, where they came to a bridge, it was to find that it had been broken down for years, and this would often mean a journey along the rugged banks perhaps for miles before they found a place where it was wise to try and ford the swollen stream.

There was always something, though, to interest the professor—a watch-tower in ruins at the corner of some defile, the remains of a castle, an aqueduct, a town with nothing visible but a few scattered stones, or a cemetery with the remains of marble tombs.

Day after day fresh ruins to inspect, with the guide proving his value more and more, and relieving the party a great deal from the pertinacious curiosity of the scattered people, who would not believe that the travellers were visiting the country from a desire for knowledge.

It must be for the buried treasures of the old people, they told Yussuf again and again; and they laughed at him derisively as he repeated his assurances.

"Don't tell them any more," Lawrence used to say in a pet; "let the stupids waste their time."

Sometimes this constant examination of old marbles and this digging out of columns or slabs grew wearisome to the lad, but not often, for there was too much exciting incident in their travels through gorge and gully—along shelves where the horses could hardly find foothold, but slipped and scrambled, with terrible precipices beneath, such as at first made the travellers giddy, but at last became so common, and their horses gave them so much confidence, that they ceased to be alarmed.

It was a wonderful country, such as they had not dreamed could exist so near Europe, while everywhere, as the investigations went on, they were impressed with the feeling that, unsafe as it was now, in the past it must have been far worse, for on all hands there were the remains of strongholds, perched upon the top of precipitous heights with the most giddy and perilous of approaches, where, once shut in, a handful of sturdy Greeks or stout Romans could have set an army at defiance. This was the more easy from the fact that ammunition was plentiful in the shape of rocks and stones, which the defenders could have sent crashing down upon their foes.

It was one evening when the difficulties of the day's journey had been unusually great that they were on their way toward a village beyond which, high up in the mountains, Yussuf spoke of a ruined city that he had only visited once, some twenty years before. He had reserved it as one of the choicest bits for his employers, and whenever Lawrence had been enraptured over some fine view or unusually grand remains Yussuf had smiled and said " Wait."

Their progress that day had been interrupted by a storm, which forced them to take shelter for a couple of hours, during which the hail had fallen in great lumps as big as walnuts, and when this was over it lay on the ridges in white beds and crunched beneath the feet of their horses.

Their way lay along one of the defiles where the road had been made to follow the edge of the stream, keeping to its windings; but as they descended a slope, and came near the little river, Yussuf drew rein.

"It is impossible, excellencies," he said; "the path is covered by the torrent, and the water is rising fast."

"But is there no other way—a mile or two round?" said the professor,

Yussuf shook his head as he pointed to the mountains that rose on every side.

"It is only here and there that there is a pass," he said. "There is no other way for three days' journey. We must go back to the place where we sheltered and wait till the river flows back to its bed."

"How long?" asked Mr. Burne; "an hour or two?"

"Perhaps longer, effendi," said Yussuf. "Mind how you turn round; there is very little room."

They had become so accustomed to ride along shelves worn and cut in the mountain sides that they had paid little heed to this one as they descended, their attention having been taken by the hail that whitened the ledges; but now, as they were turning to ascend the steep slope cut diagonally along the precipitous side of the defile, the dangerous nature of the way became evident.

But no one spoke for fear of calling the attention of his companions to the risky nature of the ride back; so, giving their horses the rein, the docile beasts planted their feet together, and turned as if upon a pivot before beginning to ascend.

So close was the wall of rock in places that the baggage brushed the side, and threatened to thrust off the horses and send them headlong down the slope, that began by being a hundred feet, and gradually increased till it was five, then ten, and then at least fifteen hundred feet above the narrow rift, where the stream rushed foaming along, sending up a dull echoing roar that seemed to quiver in the air.

How it happened no one knew. They had plodded on, reaching the highest part, with Hamed and the baggage horses in front, for there had been no room to pass them. First Yussuf, then the professor, Mr. Burne and Lawrence on Ali Baba, of course counting from the rear. There was a good deal of hail upon the path, but melting so fast in the hot sun that it was forgotten, and all were riding slowly on, when the second baggage horse must have caught its load against the rock, with the result that it nearly fell over the side. The clever beast managed to save itself, and all would have been well had it not startled Ali Baba, who made a plunge, stepped upon a heap of the hail, and slipped, the left fore-hoof gliding off the ledge.

The brave little animal made a desperate effort to recover itself, but it had lost its balance, and in its agony it made a bound, which took it ten feet forward, and along the rapid slope, where it seemed to stand for a moment, and then, to the horror of all, it began to slip and stumble rapidly down the steep side of the ravine towards a part that was nearly perpendicular, and where horse and rider must be hurled down to immediate death.

Everyone remained motionless as if changed to stone, while the clattering of the little horse's hoofs went on, and great fragments went rattling off beneath it to increase their pace and go plunging down into the abyss as if to show the way for the horse to follow to destruction.

It was all a matter of moments, with the gallant little

beast making bound after bound downward, as it felt that it could not retain its position, while Lawrence sat well back in his saddle, gripping it tightly with his knees, and holding the loosened rein.

Another bound, and another, but no foothold for the horse, and then, after one of its daring leaps, which were more those of a mountain sheep or goat than of a horse, Ali Baba alighted at the very edge of the perpendicular portion of the valley side, and those above saw him totter for a moment, and then leap right off into space.

CHAPTER XXXII.

ANOTHER SERPENT.

THE professor uttered a groan, and covered his eyes.

But only for a moment. The next he was descending from his horse, and beginning to clamber down the side of the precipice, but a cry from Yussuf stopped him.

"No, no, effendi. We must go back down to the side of the river and climb up. We cannot descend."

It was so plain that the professor said nothing; but, as if yielding to the command of a superior officer, clambered back to the pathway, and all stood gazing down to where the slope ended and the perpendicular wall began.

There was nothing to see but the top of the wall

of rock: nothing to hear but the hissing, roaring rush of the water far below.

"Come," said Yussuf, turning his horse, and taking the lead in the descent along the path they had just reascended, down which, scrambling and slipping over the thawing ice, they crept slowly, looking in the midst of the stupendous chasm little bigger than flies.

The old lawyer trembled, while the professor's cheeks looked sunken, his eyes hollow. No one spoke, and as they went on, the crunching of the half-melted hailstones and the click of the horses' hoofs against the loosened stones sounded loudly in the clear air.

It was a perilous descent, for the horses were constantly slipping; but at last the bottom of the defile was reached, and the steeds being left in charge of Hamed, Yussuf turned sharply to the right, closely followed by Mr. Preston and Mr. Burne, to climb along the steep stone-burdened slope, where the flooded mountain torrent was just beneath them and threatening to sweep them away.

Yussuf turned from time to time to look at his companions, half expecting that they would not follow, for the way he took was extremely perilous, and he fully expected to see Mr. Preston give up in despair. But, experienced as he was in the ways of Englishmen, he did not quite understand their nature, for not only was the professor toiling on over the mossy stones just behind him, but Mr. Burne, with his face glistening in perspiration and a set look of determination in his features, was clambering up and sliding down with unwonted agility, but with a piteous look in his eyes

which told how painfully he felt the position in which
they were placed.

No one spoke, every effort being needed for the
toilsome task, as they clambered along, now down in
narrow rifts, now dragging themselves painfully over
the rugged masses of rock which lay as they had fallen
from the side of the defile, a couple of thousand feet
above them. The scene would have appeared magnificent at another time; the colours of the rocks, the tufts
of verdant bushes, the gloriously-mossed stones, the
patches of white hail, and the glancing, rushing, and
gleaming torrent, which was here deep and dark, there
one sheet of white effervescing foam. But the hearts
of all were too full, and their imaginations were paint-
ing the spectacle upon which they soon expected to
gaze, namely, the terribly mutilated body of poor
Lawrence, battered by his fall out of recognition.

One moment Mr. Preston was asking himself how
he could make arrangements for taking the remains of
the poor lad home. At another he was thinking that
it would be impossible, and that he must leave him
sleeping in this far-off land. While, again, the course
of his thoughts changed, and he found himself believing that poor Lawrence would have fallen and rolled
on, and then, in company with the avalanche of loose
stones set in motion by his horse's hoofs, have been
plunged into the furious torrent, and been borne away
never to be seen again.

A curious dimness came over the professor's eyes, as
he paused for a moment or two upon the top of a rock,
to gaze before him. But there was nothing visible, for
the defile at the bottom curved and zigzagged so that

they could not see thirty yards before them, and where it was most straight the abundant foliage of the trees growing out of the cliffs rendered seeing difficult.

"It must have been somewhere here, effendi," said Yussuf at last, pausing for the others to overtake him, and pointing upwards. "Let us separate now, and search about. You, Mr. Burne, keep close down by the river; you, Mr. Preston, go forward here; and I will climb up—it is more difficult—and search there. I will shout if I have anything to say."

The professor looked up to find that he was at the foot of a mass of rock, high up on whose side there seemed to be a ledge, and then another steep ascent, broken by shelves of rock and masses which seemed to be ready to crumble down upon their heads.

Each man felt as if he ought to shout the lad's name, and ask him to give some token of his whereabouts, but no one dared open his lips for the dread of the answer to the calls being only the echoes from the rocks above, while beneath there was the dull, hurrying roar of the torrent which rose and fell, seeming to fill the air with a curious hissing sound, and making the earth vibrate beneath their feet.

They were separating, with the tension of pain upon their minds seeming more than they could bear, when, all at once, from far above, there was a cry which made them start and gaze upward.

"Ahoy—y—y!"

There was nothing visible, and they remained perfectly silent—listening, and feeling that they must have been mistaken; but just then a stone came bounding down, to fall some fifty feet in front, right on to a mass of rock, and split into a score of fragments.

Then again:

"Ahoy! Where are you all?"

"Lawrence, ahoy!" shouted the professor, with his hands to his mouth.

"Ahoy!" came again from directly overhead. "Here. How am I to get down?"

All started back as far as they could to gaze upward, and then remained silent, too much overcome by their emotion to speak, for there, perched up at least a thousand feet above them, stood Lawrence in an opening among the trees, right upon a shelf of rock. They could see his horse's head beside him, and the feeling of awe and wonder at the escape had an effect upon the party below as if they had been stunned.

"How—am—I—to—get—down?" shouted Lawrence again.

Yussuf started out of his trance and answered:

"Stay where you are. I will try and climb up."

"All right," cried Lawrence from his eyrie.

"Are you hurt, my boy?" cried Mr. Preston; and his voice was repeated from the face of the rock on the other side.

"No, not much," came back faintly, for the boy's voice was lost in the immensity of the place around.

"We will come to you," cried the professor, and he began to follow Yussuf, who was going forward to find the end of the mass of rock wall, and try to discover some way of reaching the shelf where the boy was standing with his horse.

"Are you coming too, effendi?" said Yussuf at the end of a few minutes' walking.

"Yes," said the professor. "You will wait here, will you not, Burne?"

"Of course I shall—not," said the old lawyer. "You don't suppose that I am going to stand still and not make any effort to help the boy, do you, Preston? Hang it all, sir! he is as much interest to me as to you."

It was evident that Mr. Burne was suffering from exhaustion, but he would not give in, and for the next two hours he clambered on after his companions, till it seemed hopeless to attempt farther progress along the defile in that direction, and they were about to go back in the other, to try and find a way up there, when Yussuf, who was ahead, suddenly turned a corner and uttered a cry of delight which brought his companions to his side.

There was nothing very attractive to see when they reached him, only a rushing little torrent at the bottom of a rift hurrying to join the stream below; but it was full of moment to Yussuf, for it led upward, and it was a break in the great wall of rock.

Yussuf explained this clearly, and, plunging down, he was in a few minutes holding out his hand to his companions, and pointing out that the path was easier a few yards on.

So it proved, for the stream grew less, and they were able to climb up its bed with ease, finding, too, that it led in the direction they wanted to take, as well as upward, till, at the end of an hour, they were able to turn off along a steep slope with a wall of rock above them and another below.

The obstacles they met with were plentiful enough, but not great; and at last, when they felt that they were fully a thousand feet above the torrent, and somewhere near the spot on which they had hailed Lawrence, Yussuf stopped, but no one was to be seen.

That must be the shelf below us yonder, effendi," said the guide. " I seem to know it because of the big tree across the valley. Yes; that must be the shelf."

He led the way to try and descend to it, but that proved impossible, though it was only some fifty feet below.

Retracing their steps they were still defeated, but, upon going forward once more, Yussuf found what was quite a crack in the rocks, some huge earthquake split which proved to be passable, in spite of the bushes and stones with which it was choked, and after a struggle they found themselves upon an extensive ledge of the mountain, but no Lawrence.

" The wrong place, Yussuf," said the professor, as Mr. Burne seated himself, panting, upon a block of stone, and wiped his face.

" No, effendi; but I am sure it was here," said the Turk quietly. " Hush! what is that?"

The sound came from beyond a mass of rock, which projected from the shelf over the edge of the precipice, the perpendicular rock seeming to fall from here sheer to the torrent, that looked small and silvery now from where they stood.

" It is a horse feeding," said Yussuf smiling. "They are over yonder."

The next minute they were by the projecting rock which cut the shelf in two.

Yussuf went close to the edge, rested his hand upon the stone, and peered over.

"Only a bird could get round there," he said, shaking his head, and going to the slope above the ledge. "We must climb over."

Mr. Burne looked up at the place where they were expected to climb with a lugubrious expression of countenance; but he jumped up directly, quite willing to make the attempt, and followed his companions.

The climb proved less difficult than it seemed, and on reaching the top, some fifty feet above where they had previously stood, there below them stood Ali Baba, cropping the tender shoots of a large bush, and as soon as he caught sight of them he set up a loud neigh.

There was no sign of Lawrence, though, until they had descended to the shelf on that side, when they found him lying upon the short growth fast asleep, evidently tired out with waiting.

"My dear boy!" was on the professor's lips; and he was about to start forward, but Yussuf caught him roughly by the shoulder, and held him back.

"Hist! Look!" he whispered.

Both the professor and Mr. Burne stood chilled to the heart, for they could see the head of an ugly gray coarsely scaled viper raised above its coil, and gazing at them threateningly, after having been evidently alarmed by the noise which they had made.

The little serpent had settled itself upon the lad's bare throat, and a reckless movement upon the part of the spectators, a hasty waking on the sleeper's part might end in a venomous bite from the awakened beast.

"What shall we do, Yussuf?" whispered the professor in a hoarse whisper. "I dare not fire."

"Be silent, effendi, and leave it to me," was whispered back; and, while the two Englishmen looked on with their hearts beating anxiously, the Turk slowly advanced, taking the attention of the serpent more and more.

As he approached, the venomous little creature crept from the boy's neck on to his chest, and there paused, waving its head to and fro, and menacingly thrusting out its forked tongue.

The danger to be apprehended was a movement upon the part of Lawrence, who appeared to be sleeping soundly, but who might at any moment awaken. Yussuf, however, was ready to meet the emergency, for he slowly continued to advance with his staff thrown back and held ready to strike, while, as he came nearer, the serpent seemed to accept the challenge, and crawled slowly forward, till it was upon a level with the lad's hips.

That was near enough for Yussuf, who noted how Lawrence's hands were well out of danger, being beneath his head.

He hesitated no longer, but advanced quickly, his companions watching his movements with the most intense interest, till the serpent raised itself higher, threw back its head, and seemed about to throw itself upon its advancing enemy.

The rest was done in a flash, for there was a loud *whizz* in the air as Yussuf's staff swept over Lawrence, striking the serpent, rapid as was its action, low down in the body, and the virulent little creature, broken and helpless, was driven over the edge of the precipice to fall far away among the bushes below.

"Hallo! what's that?" cried Lawrence, starting up. "Oh, you've got here, then."

"Yes; we are here, my lad," cried the professor, catching one hand, as the old lawyer took the other. "Are you much hurt?"

"Only stiff and shaken. Ali made such a tremendous leap—I don't know how far it was; and then he came down like an india-rubber ball, and bounded again and again till he could find good foothold, and then we slipped slowly till we could stop here, and it seemed as if we could go no farther."

"What an escape!" muttered Mr. Burne, looking up.

"Oh, it wasn't there," said Lawrence patting his little horse's neck. "It must have been quite a quarter of a mile from here. But how did you come?"

Yussuf explained, and then Mr. Preston looked aghast at the rock they had climbed over.

"Why, we shall have to leave the pony," he said.

"Oh, no, effendi," replied Yussuf; "leave him to me. He can climb like a goat."

And so it proved, for the brave little beast, as soon as it was led to the task by the rein passed over its head, climbed after Yussuf, and in fact showed itself the better mountaineer of the two, while, after the rock was surmounted, and a descent made upon the other side, it followed its master in the arduous walk, slipping and gliding down the torrent-bed when they reached it, till at last they reached the greater stream, which to their delight had fallen to its regular summer volume, the effects of the storm having passed away, and the sandy bed being nearly bare.

Theirs proved quite an easy task now, in spite of weariness; and as evening fell, they reached Hamed, camped by the roadside, with the horses grazing on the bushes and herbage, all being ready to salute Ali Baba with a friendly neigh.

They had a long journey before them still; but there was only one thing to be done now—unpack the provisions, light a fire, make coffee, and try to restore some of their vigour exhausted by so many hours of toil.

CHAPTER XXXIII.

A FORMIDABLE PARTY.

ORTUNATELY for the travellers a glorious moonlight night followed the glowing evening, and they reached in safety a mountain village, where, awed by their appearance and display of arms, the rather surly people found them a resting-place.

For days and days after this their way was on and on, among the mountains, deeper and deeper in the grand wild country. Sometimes they encountered good and sometimes surly treatment, but the beauty of the scenery and the wonderful remains of ancient occupation recompensed the professor, while Mr. Burne in his snappish manner seemed to be satisfied in seeing

Lawrence's interest in everything around him, his relish for the various objects increasing every hour.

For the change was unmistakable; he was making rapid progress back to health; and instead of the rough life and privations of hunger, thirst, and exposure having a bad effect, they seemed to rouse up in his nature a determination that rapidly resulted in vigour.

" What are you going to show us to-day, Yussuf?" asked the lad, one glorious autumn morning, when the little party were winding along one of the many mountain tracks, so like others they had passed that they might have been repeating their journey.

" Before long we shall reach the great ruins of which I have so often spoken," replied Yussuf, smiling at the boy's eager look.

" At last!" cried Lawrence. " I began to think that we were never going to get there. But is there nothing to see to-day?"

" Yes," replied Yussuf. " We are approaching a village now. It lies yonder low down in this rift— where the cedars are half-way up on that shelf in the mountain side."

" Yes; I see," replied Lawrence; " but what a place! Why, they must be without sun half their time."

" Oh, no, effendi," said Yussuf; " certainly they are in shadow at times, but though the village seems to lie low, we are high up in the mountains, and when it is scorching in the plains, and the grass withers for want of water, and down near the sea people die of fever and sunstroke, up here it is cool and pleasant, and the flowers are blossoming, and the people gather in their fruit and tend their bees."

"And in the winter, Yussuf?" said the professor, who had been listening to the conversation.

"Ah, yes, in the winter, effendi, it is cold. There is the snow, and the wolves and the bears come down from the mountains. It is a bad time then. But what will you?—is it always summer and sunshine everywhere? Ah! look, effendi Lawrence," he cried, pointing across the narrow gorge, "you can see from here."

"See what?" cried Lawrence. "I can only see some holes."

"Yes; those are the caves where the people her keep their bees. The hives are in yonder."

"What, in those caves?"

"Yes; the people are great keepers of bees, for they thrive well, and there is abundance of blossom for the making of honey."

"But why do they put the hives in yonder?"

"In the caves? Because they are out of the sun, which would make the honey pour down and run out in the hot summer time, and in the winter the caverns are not so cold. It does not freeze hard there, and the hives are away out of the snow, which lies so heavy here in the mountains. It is very beautiful up here, and in the spring among the trees there is no such place anywhere in the country for nightingales; they fill the whole valley with their song. Now, effendi, look before you."

They had reached a turn in the valley, where once more a grand view of the mountain chain spread before them, far as eye could reach, purple mountains, and beyond them mountains that seemed to be of silver, where the snow capped their summits.

But among them were several whose regular form took the professor's attention directly, and he pointed them out.

"Old volcanoes," he said quietly.

"Where?" cried Lawrence. "I want above all things to see a burning mountain."

"You can see mountains that once burned," said the professor; "but there are none here burning now."

"How disappointing!" cried Lawrence. "I should like to see one burn."

"Then we must go and see Vesuvius," cried Mr. Burne decisively. "He shall not be disappointed."

"I think the young effendi may perhaps see one burning a little here," said Yussuf quietly. "There are times when a curious light is seen floating up high among the mountains. The peasants call it a spirit light, but it must be the sulphurous glare rising from one of the old cones, above some of which I have seen smoke hanging at times."

"Why, Yussuf, you are quite a professor yourself, with your cones, and sulphurous, and arguments," cried Mr. Burne.

"A man cannot be wandering all his life among nature's wonders, effendi, and showing English, and French, and German men of wisdom the way, without learning something. But I will watch each night and see if I can make out the light over the mountains."

"Do, Yussuf," cried the professor eagerly.

Yussuf bowed.

"I will, excellency, but it is not often seen—only now and then."

They began to descend the side of the defile, and

before long came upon a fine grove of ancient planes, upon some of whose leafless limbs tall long-necked storks were standing, placidly gazing down at them unmoved; and it was not until the party were close by that they spread their wings, gave a kind of bound, and floated off, the protection accorded to them making them fearless in the extreme.

"Stop!" cried the professor suddenly, and the little party came to a stand by a rough craggy portion of the way where many stones lay bare.

"Well, what is it?" cried Mr. Burne impatiently. "I'm sure those are natural or live stones, as you call them."

"Yes," said the professor; "it was not the stones which attracted me, but the spring."

"Well, we have passed hundreds of better springs than that, and besides it is bad water; see, my horse will not touch it."

"I thought I was right," cried the professor dismounting. "Look here, Lawrence, that decides it; here is our first hot spring."

"Hot?" cried Lawrence, leaping off and bending over the spring. "Why, so it is."

"Yes, a pretty good heat. This is interesting."

"It is a volcanic country, then," said Lawrence eagerly. "Oh, Mr. Preston, we must see a burning mountain."

"It does not follow that there are burning mountains now," said the professor smiling, "because we find hot springs."

"Doesn't it?" said Lawrence in a disappointed tone.

"Certainly not. You would be puzzled to find a

volcano in England, and yet you have hot springs in Bath."

"Effendi, be on your guard. I do not like the look of these people," said Yussuf quickly, for a party of mounted men, all well armed, was seen coming from the opposite direction; but they passed on scowling, and examining the little group by the hot spring suspiciously.

"A false alarm, Yussuf," said the professor smiling.

"No, effendi," he replied; "these are evil men. Let us get on and not stop at this village, but make our way to the next by another track which I know, so as to reach the old ruined city, and they may not follow. If they do, I think they will not suspect the way we have gone."

There seemed to be reasons for Yussuf's suspicions, the men having a peculiarly evil aspect. A perfectly honest man sometimes belies his looks, but when a dozen or so of individuals mounted upon shabby Turkish ponies, all well armed, and wearing an eager sinister look upon their countenances, are seen together, if they are suspected of being a dishonest lot, there is every excuse for those who suspect them.

"'Pon my word, Preston," said Mr. Burne, "I think we had better get off as soon as possible."

"Oh, I don't know," replied the professor; "the men cannot help their looks. We must not think everyone we see is a brigand."

"You may think that those are, effendi," said Yussuf in his quiet way. "Let us get on. You go to the front and follow the track beyond the village—you can make no mistake, and I will hang back and try and find out whether we are followed."

"Do you think there is danger, then?" whispered the professor.

"I cannot say, effendi; it may be so. If you hear me fire, be on your guard, and if I do not return to you, hasten on to the next village, and stay till you have sent messengers to find an escort to take you back."

"Yussuf! is it so serious as that?"

"I don't know, effendi. I hope not, but we must be prepared."

CHAPTER XXXIV.

A STARTLING CHECK.

USSUF'S suspicions seemed to be without reason, for the rest of that day's journey was finished without adventure, and the party reached a village and found good quarters for the night.

So comfortable were they that the scare was laughed at, and it seemed to all three that Yussuf was rather ashamed of his timidity.

Contrary to their experience of many nights past they found the headman of the village civil and even humble; but it did not excite the suspicion of the travellers, who congratulated themselves upon their good fortune.

The only drawback to their comfort was the fact

that Lawrence was suffering somewhat from the shock
of his descent from the rocky shelf.

At first he had merely felt a little stiff, the excitement of the whole adventure tending to keep his
thoughts from his personal discomfort; but by degrees
he found that he had received a peculiar jar of the
whole system, which made the recumbent position the
most comfortable that he could occupy.

It was no wonder, for the leaps which the pony had
made were tremendous, and it was as remarkable that
the little animal had kept its feet as that Lawrence
had retained his seat in the saddle.

The next morning, a memorable one in their journey,
broke bright and clear; and Lawrence, after a hearty
breakfast of bread, yaourt, and honey, supplemented
by coffee which might have been better, and peaches
which could not have been excelled, mounted Ali Baba
in the highest of spirits, feeling as he did far better
for his night's rest. The sun was shining gloriously
and lighting up the sides of the mountains and flashing from the streams that trickled down their sides.
Low down in the deep defiles there were hanging
mists which looked like veils of silver decked with
opalescent tints of the most delicate transparency, as
they floated slowly before the morning breeze.

Their host of the night wished them good speed
with a smiling face, and they were riding off when
Lawrence happened to look back and saw that the
man had taken off his turban and was making a derisive gesture, to the great delight of the group of people
who were gathered round.

Lawrence thought it beneath his notice and turned

away, but this once more seemed to give strength to
Yussuf's suspicions.

But a bright morning in the midst of the exhilarat-
ing mountain air is not a time for bearing in mind
suspicions, or thinking of anything but the beauty
of all around. They were higher up in the mountains
now, with more rugged scenery and grand pine-woods;
and as they rode along another of the curious shelf-
like tracks by the defile there was constantly something fresh to see.

They had not been an hour on the road before
Yussuf stopped to point across the gorge to an object
which had taken his attention on the other side.

"Do you see, effendi Lawrence?" he said smiling.

"No."

"Yonder, just to the left of that patch of bushes
where the stone looks gray?"

"Oh, yes, I see now," cried the lad—"a black sheep."

"Look again," said Yussuf; and he clapped his hands
to his mouth and uttered a tremendous "Ha-ha!"

As the shout ran echoing along the gorge the animal
on the farther slope, quite two hundred yards away,
went shuffling along at a clumsy trot for some little
distance, and then stopped and stood up on its hind-
legs and stared at them.

"A curious sheep, Lawrence!" said Mr. Preston, ad-
justing his glass; "what do you make of it now?"

"Why, it can't be a bear, is it?" cried Lawrence
eagerly.

"Undoubtedly, and a very fine one," said Mr. Preston.

"Let's have a look," said Mr. Burne; and he too
focussed his glass. "Why, so it is!" he cried—"just

such a one as we used to have upon the pomatum pots. Now, from what gardens can he have escaped?"

The professor burst out laughing merrily.

"It is the real wild animal in his native state, Burne," he said.

"Then let's shoot him and take home his skin," cried Lawrence, preparing to fire.

"You could not kill it at this distance, effendi," said Yussuf; "and even if you could, it would be a day's journey to get round to that side and secure the skin. Look!"

The chance to fire was gone as he spoke, for the bear dropped down on all-fours, made clumsily for a pile of rocks, and Mr. Preston with his glass saw the animal disappear in a hole that was probably his cave.

"Gone, Lawrence!" said the professor. "Let's get on."

"I should have liked to go on after him," said Lawrence, gazing at the hole in the rocks wistfully; "there's something so strange in seeing a real bear alive on the mountains."

"Perhaps we shall see more yet," said Yussuf, "for we are going into the wildest part we have yet visited. Keep a good look-out high up on each side, and I daresay we shall not go far without finding something."

"Right, Yussuf," cried the professor; "there is another of those grand old watch-towers. Look, Burne! —just like the others we have seen planted at the corner where two defiles meet."

"Ah, to be sure—yes," said the old lawyer. "What! an eagle's nest?"

"And there goes the eagle," cried Lawrence, pointing,

as a huge bird swept by them high up on rigid wing, seeming to glide here and there without the slightest effort. "That's an eagle, is it not, Mr. Preston?"

"A very near relative, I should say," replied the professor. "The lammergeier, as they call it in the Alpine regions. Yes, it must be. What a magnificent bird!"

"We shall see more and finer ones, I daresay," said Yussuf quietly; "but the time is passing, excellencies. We have a long journey before us, and I should like to see the better half of a difficult way mastered before mid-day."

Their guide's advice was always so good that they continued their slow progress, the baggage horses ruling the rate at which they were able to proceed; and for the next hour they went on ascending and zigzagging along the rugged mountain track, with defile and gorge and ridge of rock rising fold upon fold, making their path increase in grandeur at every turn, till they were in one of nature's wildest fastnesses, and with the air perceptibly brisker and more keen.

All at once, just as they had turned into the entrance to one of the most savage-looking defiles they had yet seen, Yussuf pointed to a distant pile of rock and said sharply:

"Look, there is an animal you may journey for days without seeing. Take the glass, effendi Lawrence, and say what it is."

The lad checked his pony, adjusted his glass, an example followed by the professor, while Mr. Burne indulged himself with a pinch of snuff.

"A goat," cried Lawrence, as he got tne animal into

the field of the glass, and saw it standing erect upon
the summit of the rock, and gazing away from them
—a goat! And what fine horns?"

"An ibex, Lawrence, my boy. Goatlike if you like.
Ah, there he goes. How easily they take alarm."

For the animal made a bound and seemed to plunge
from rock to rock down into a rift, and then up an
almost perpendicular wall on the opposite side higher
and higher until it disappeared.

"It is no wonder, excellency," said Yussuf as they
rode on along the narrow path, "when every hand is
against them, and they have been taught that they are
not safe from bullets half a mile away, and—Why is
Hamed stopping?"

They had been halting to gaze at the ibex, and all
such pauses in their journey were utilized for letting
Hamed get well on ahead with his slow charge. Experience had taught them that to leave him behind with
the necessaries of life was often to miss them altogether till the next morning.

In this case he had got several hundred yards in
advance, but had suddenly stopped short, just at the
point of a sharp elbow in the track, where they could
see him with the two horses standing stock-still, and
staring straight before him.

"Let's get on and see," said the professor, and they
pressed on to come upon a spot where the track forked
directly after, a narrower path leading up a rift in the
mountains away to their left, and the sight of this
satisfied Yussuf.

"Hamed thinks he may be doing wrong," he said,
"and that perhaps he ought to have turned down

here.   All right, go on!" he shouted in his own tongue,
as they rode on past the wild passage among the rocks.

But Hamed did not stir, and as they advanced they
could see that he was sheltering himself behind one of
his horses, and still staring before him.

The way curved in, and then went out to the
shoulder upon which the baggage horses stood, doubt-
less bending in again directly on the other side. Hence,
then, it was impossible for Yussuf and his party to see
what was beyond; neither could they gain a sight
by altering their course, for their path was but a shelf
with the nearly perpendicular side of the gorge above
and below.

They were now some eighty or ninety yards from
the corner, and Yussuf shouted again:

"Go on, man; that is right."

But Hamed did not move hand or foot, and Yussuf
checked his horse.

" There is something wrong, effendis," he said quietly;
and he thrust his hand into his breast and drew out
his revolver.   "Get your weapons ready."

" What, is there to be a fight?" said Mr. Burne excitedly.

" I hope not," said Mr. Preston gravely, as he examined the charge of his double gun, an example
followed by Lawrence, whose heart began to beat
heavily.

" You had better halt here, excellencies," said Yussuf.
" I will go forward and see."

" No," said Mr. Preston; " we will keep together.
It is a time for mutual support.   What do you think
it is?"

"The man is timid," said Yussuf. "He is a good driver of horses, but a little frightens him. The country is wild here; there may be wolves or a bear on the track which he would not dare to face, though they would run from him if he did."

They all advanced together with their weapons ready for immediate use, and Lawrence's hands trembled with eagerness, as he strained his eyes forward in expectation of a glimpse at bear or wolf, and in the hope of getting a good shot.

"Why don't you speak? Are you ill?" continued Yussuf as he rode on forward. But Hamed did not stir; and it was not until the guide could almost touch him that he was able to see what was the cause of his alarm, and almost at the same moment the others saw it too.

"We must keep a bold face and retreat," said Yussuf in a quick low tone. "You, Hamed, take the bridle of that horse and lead him back; the other will follow."

"No, no, no; they will fire.'

"So shall I," said Yussuf, placing the muzzle of his pistol close to the man's ear. "Obey me; or—"

Hamed shuddered and began to implore, but Yussuf was rigid.

"Go on back," he said forcing himself round the foremost horse, closely followed by the professor, though there was hardly room for their steeds to pass, and there was a fall of several hundred feet below, while, pressed like this, Hamed began to whimper; but he obeyed, and led the horses past Lawrence and Mr. Burne, who now went forward, eager and excited to know what was wrong, and upon joining their companions it was to find themselves face to face with a gang of about twenty fierce-looking men, all mounted, and who were seated with their guns presented toward the travellers' heads.

CHAPTER XXXV.

BROUGHT TO BAY.

THE strangers were some fifty yards away, and thoroughly blocked all further progress. What they were was not open to doubt; but, though they sat there presenting their guns, they did not attempt to fire, nor yet to advance, contenting themselves by barring the travellers' way.

"Do you think they are enemies, Yussuf?" said Mr. Preston calmly.

"There is no doubt of it, effendi," was the reply.

"But had we not better ride boldly forward? They will not dare to stop us. Besides, if they do, we are well armed."

"They are twenty and we are only two, effendi, for we cannot depend upon three of our party. It would be no use to attack. We must retreat steadily, and get back to the village; they will not dare to follow us so far."

"What do you propose doing, then?"

"For one of us to remain here facing them, till the others have got fifty yards back. Then one is to turn and face the scoundrels till I have ridden in, and on with the others another fifty yards or so, when I face round, and the one on duty rides in, and so on by turns. If we keep a bold front we may hold them off."

" A good plan," said the professor; "but would it not be better for two to face them, and two to go forward —I mean, to retreat?"

" No, effendi; there is not too much room for the horses. Do as I ask."

Mr. Preston obeyed on the instant, and with Hamed in front the retreat was commenced, all retiring and leaving Yussuf on the projecting corner, weapon in hand, and a sword hanging from his wrist by the knot.

Then, at about fifty yards, Mr. Preston halted and faced round, with gun presented, and as the others still rode on, Yussuf left his post and joined the professor, passing him and riding on another fifty yards behind, where he faced round in turn.

As the professor made his horse face about and rode on, he had only just reached the guide, when a clattering of horses' hoofs behind him made him look sharply round.

The enemy had advanced, and about half a dozen men had taken up the vacated position at the elbow of the track.

There they stopped, looking menacing enough, but making no advance, merely watching the progress of the little party as they retreated round the curve towards the other corner which they had passed on their way.

"Had we not better get on faster?" said the professor.

"No," replied Yussuf; "we must go slowly, or they will close in; and your excellency does not want blood to be shed. Our only chance is by keeping a bold front, and retreating till we can get help. They will not dare to attack us if we keep on like this, for they do not care to risk their lives."

"Go on then," said the professor; and the retreat was kept up for about ten minutes, and then came to a stop, for Hamed, on reaching the other corner with his baggage horses, stopped short suddenly, and on Lawrence trotting up to him, the professor saw him too stop, and present his gun.

" We are trapped, effendi," said Yussuf sadly.

"Trapped!" cried Mr. Preston sharply. " What do you mean?"

" The dogs have another party who have been hidden in that side track, and they have come out as soon as we passed. We are between two fires. What shall we do?"

It was plain enough, for the next minute Hamed and Lawrence were seen to be driven back, and a party similar to that upon the first corner stood out clearly in the morning air—a gang before, and one behind, and the precipice above and below. It was either fight or yield now, and Yussuf had asked the question, what was to be done.

Shut in as they were completely, the little party closed up together on the curved path, Hamed requiring no telling, while the enemy made no attempt to advance.

Mr. Burne took out his box, had a large pinch of

snuff, and then blew his nose so outrageously that the horses pricked their ears, and Ali Baba snorted and looked as if he would try another of his wonderful leaps if that kind of thing were to be continued.

"Well, Yussuf," said the professor, "what is to be done?"

The guide sighed deeply and looked full in his employer's face.

"Excellency," he said softly, "I feel as if all my bones were turned to water."

"Oh, indeed, sir," cried Mr. Burne sharply; "then you had better turn them back to what they were."

"What is to be done, Yussuf?" continued the professor. "If we make a stout resistance, shall we beat them off?"

"No, effendi," said Yussuf sadly; "it is impossible. We might kill several, but they are many, and those who are left do not value life. Besides, effendi, some of us must fall."

"What are these people, then?"

"Brigands—robbers, excellency."

"Brigands and robbers in the nineteenth century!" cried Mr. Burne angrily; "it is absurd."

"In your country, excellency; but here they are as common as they are in Greece."

"But the law, sir, the law!" cried Mr. Burne. "Confound the scoundrels! where are the police?"

Yussuf shrugged his shoulders.

"We are far beyond the reach of the law or the police, excellency, unless a little army of soldiers is sent to take or destroy these people; and even then what can they do in these terrible fastnesses, where the

brigands have hiding-places and strongholds that cannot be found out, or if found, where they can set the
soldiery at defiance?"

Mr. Burne blew his nose again fearfully, and created
a series of echoes that sounded as if old men were blowing their noses from where they stood right away to
Constantinople, so strangely the sounds died away in
the distance.

"Then why, sir, in the name of common sense and
common law, did you bring us into this out-of-the-way
place, among these dirty, ragged, unshaven scoundrels?
It is abominable! It is disgraceful! It is—"

"Hush! hush! Burne; be reasonable," said the professor. "Yussuf has only obeyed orders. If anyone is
to blame it is I, for I wished to see this ruined fast-
ness of the old Roman days."

Yussuf smiled, and gave the professor a grateful
look.

"Humph! It's all very well for you to take his
part. He ought to have known," grumbled the old
lawyer.

"Travellers are never free from risk in any of the
out-of-the-way parts of the country," said Yussuf
quietly.

"And of course we knew that, and accepted the
risk," said the professor. "Come, come, Burne, be reasonable. Yussuf is not to blame. The question is,
What are we to do—fight or give up?"

"Fight," said Mr. Burne fiercely. "Hang it all, sir!
I never give in to an opponent. I always say to a
client, if he has right upon his side, 'Fight, sir, fight.'
And that's what I'm going to do."

"Fight, eh?" said the professor gravely.

"Yes, sir, fight, and I only wish I understood the use of this gun and long knife as well as I do that of a ruler and a pen."

"Look here, Yussuf, if we fight, what will be the consequences?"

"I will fight for your excellencies to the last," said the Turk calmly; "but I am afraid that we can do no good."

"Confound you, sir!" cried Mr. Burne; "if we give in they will take off our heads."

"No, no, excellency, they will make us prisoners, and strip us of our arms and all that we have of value."

"Humph! Is that all?"

"No, excellency. They will demand a heavy ransom for your release—so many Turkish pounds."

"Then we'll fight," cried Mr. Burne furiously. "I never would and I never will be swindled. Ransom indeed! Why, confound it all, Preston! is this real, or is it a cock-and-bull story told in a book?"

"It is reality, Burne, sure enough," said the professor calmly; "and I feel with you, that I would sooner fight than give up a shilling; but, cowardly as it may seem, I fear that we must give up."

"Give up? Never, sir. I am an Englishman," cried the old lawyer.

"But look at our position. We are completely at their mercy. Here we are in the centre of this half-moon curve, and the scoundrels hold the two horns in force."

"Then we'll dash up the mountain."

"It is impossible, excellency," said Yussuf.

T

"Then we'll go downwards."

"To death, Burne?" said the professor smiling.

"Confound it all!" cried Mr. Burne, "we are in a complete trap. Here, you, Yussuf, this is your doing, and you are in league with these rascals to rob us."

"Excellency!"

"Oh, Mr. Burne!" cried Lawrence, with his face scarlet; and he leaned towards Yussuf, and held out his hand to the Turk, who sat with angry, lowering countenance upon his horse.

"Mr. Burne is angry, Yussuf," said the professor in a quiet, stern manner. "He does not mean what he says, and I am sure he will apologize as an English gentleman should."

Yussuf bowed coldly, and Mr. Preston continued:

"I have the most perfect confidence in your integrity, sir, and as we are brothers in misfortune, and you know these people better than we—"

"Of course," said Mr. Burne, with an angry ejaculation.

"I ask you," said Mr. Preston, "to give us your advice. What had we better do—fight or give up?"

Yussuf's face brightened, and he turned to the old lawyer.

"Effendi," he said gravely, "you will know me better before we part, and you will tell me you are sorry for what you have said."

"I won't, sir! No, confound me, never!" cried the old lawyer; and he blew his nose like a challenge upon a trumpet.

"I am deeply grieved, effendi," continued Yussuf, smiling as he turned to the professor, "for this is a terrible misfortune, and you will be disappointed of your visit to the old city. But it would be madness to fight. We should be throwing away our lives, and that of the young effendi here, who has shown us of late that he has a long and useful life to lead. It is our fate. We must give up."

"Never!" cried Mr. Burne, cocking his gun.

"Don't be foolish, my dear Burne," said the professor. "I would say, let us fight like men; but what can we do against fifty well-armed scoundrels, who can shelter themselves and pick us off at their ease? Come, keep that gun still, or you will shoot one of us instead of an enemy."

"Now, that's cruel!" cried Mr. Burne with an air of comical vexation. "Well, I suppose you are right. Here, Yussuf, old fellow, I beg your pardon. I was only in a savage temper. I suppose we must give in; but before I'll pay a shilling of ransom they shall take off my head."

Yussuf smiled.

"Confound you, sir, don't grin at a man when he's down," cried Mr. Burne. "You've got the better of me, but you need not rejoice like that."

"I do not rejoice, excellency, only that you believe in me once more."

"Here! hi! you black-muzzled, unbelieving scoundrels, leave off, will you! Don't point your guns at us, or, by George and the dragon and the other champions of Christendom, I will fight."

He had looked at the two points of the half-moon road, and seen that about a dozen men were now dismounted, and were apparently taking aim at them.

"Well, Yussuf, we give up," said the professor. "Perhaps, after all, they may be honest people. Will you go to them and ask what they want with us?"

"They are brigands, excellency."

"Well, ask them what they will take to let us continue our journey in peace," cried Mr. Burne. "Offer 'em five shillings all round; I suppose there are about fifty—or, no, say we will give them ten pounds to go about their business; and a precious good day's work for the ragged jacks."

"I will go forward," said Yussuf. "Excellency," he continued to the professor, "trust me, and I will make the best bargain I can."

"Go on, then," said the professor; "but is there any risk to yourself?"

"Oh, no, effendi, none at all. I have no fear. They will know I come as an ambassador."

"Go on, then," said the professor; and the Turk rode slowly forward to the men, who blocked their way, and who still held their guns menacingly before them as if about to fire.

CHAPTER XXXVI.

GOOD OUT OF EVIL.

"WE'VE brought our pigs to a pretty market," grumbled Mr. Burne, as they sat watching Yussuf ride up to the brigands. "It means ruin, sir, ruin."

"There's no help for it, Burne," said the professor calmly; "it is of no use to complain."

"I am an Englishman, sir, and I shall grumble as much and as long as I please," cried the old gentleman snappishly: "and you, Lawrence, if you laugh at me, sir, I'll knock you off your horse. Here, what was the use of our buying weapons of war, if we are not going to use them?"

"Their conversation has been short," said the professor. "I suppose it is settled. So vexatious too, when we were quite near the ancient stronghold."

"Bah! you've seen old stones and ruins enough, man. I wish to goodness we were back in London. Well, Yussuf, what do they say?"

"That if your excellencies will surrender peaceably, you shall not be hurt. There is nothing else for us to do but give up."

"And you advise it, Yussuf?" said the professor.

"Yes, your excellency, we must give up; and perhaps if you are patient I may find a means for us to escape."

"Hah! that's better," cried Mr. Burne; "now you are speaking like a man. Come along, then, and let's get it over. Can the brutes speak English?"

"No, excellency, I think not. Shall I lead?"

"No," said Mr. Burne. "I shall go first, just to show the miserable ruffians that we are not afraid of them if we do give up. Come along, Preston. Confound them! how I do hate thieves."

He took a pinch of snuff and then rode slowly on with an angry contemptuous look, closely followed by his companions, to where the brigands were awaiting them with guns presented ready to fire if there was any resistance.

As they advanced, the party behind closed up quickly, all being in the same state of readiness with their weapons till the travellers found themselves completely hemmed in by as evil-looking a body of scoundrels as could be conceived. Every man had his belt stuck full of knives and pistols, and carried a dangerous-looking gun—that is to say, a piece that was risky to both enemy and friend.

One man, who seemed to hold pre-eminence from the fact that he was half a head taller than his companions, said a few words in a sharp fierce manner, and Yussuf spoke.

"The captain says we are to give up all our arms," he said; and the professor handed his gun and sword without a word, the appearance of the weapons apparently giving great satisfaction to the chief.

"Here, take 'em," growled Mr. Burne; "you ugly-looking unwashed animal. I hope the gun will go off of itself, and shoot you. I say, Preston, you haven't given them your revolver."

"Hush! neither am I going to without they ask for it. Yussuf is keeping his."

"Oh, I see," said the old lawyer brightening.

Lawrence had to resign his handsome gun and sword next, the beauty of their workmanship causing quite a buzz of excitement.

After this, as Lawrence sat suffering a bitter pang at losing his treasured weapons, the chief said a few words to Yussuf.

"The captain says, excellencies, that if you will ride quietly to their place, he will not have you bound. I have said that you will go."

"Yes," said the professor, "we will go quietly."

The chief seemed satisfied, and the prisoners being placed in the middle, the whole band went off along the mountain path, higher and higher hour after hour.

There was no attempt made to separate them, nor yet to hinder their conversation; and the brigands seemed less ferocious now that the business of the day had had so satisfactory a finish, for they were congratulating themselves upon having made a very valuable haul, and the captives, after a time, began to look upon their seizure as more interesting and novel than troublesome. That is to say, all but the professor, who bemoaned bitterly the fact that he should miss seeing the old ruined stronghold in the mountains, which was said to be the highest ruin in the land.

"It seems so vexatious, Yussuf," he said towards evening, after a very long and tedious ride through scenery that was wild and grand in the extreme; "just, too, as we were so near the aim of all my desire."

"Bother!" said Mr. Burne, "I wish they would stop and cook some dinner. Are they going to starve us?"

"No, excellency; and before an hour has passed, if I think rightly, we shall have reached the brigands' stronghold. They will not starve you, but you will have to pay dearly for all you have."

"I don't care," said Mr. Burne recklessly. "I'd give a five-pound note now for a chop, and a sovereign a-piece for mealy potatoes. This mountain air makes me ravenous, and ugh! how cold it is."

"We are so high up, excellency," said Yussuf; and then smiling, "Yes, I am right."

"What do you mean?" said the professor.

"I did not like to speak before, effendi," he said excitedly, "for I was not sure; but it is as I thought; they have now turned into the right road. Everything points to it."

"Look here," grumbled Mr. Burne, "I'm not in a humour to guess conundrums and charades; speak out, man. What do you mean?"

"I mean, excellency, that I have been wondering where the brigands' strong place could be, and I believe I have found out"

"Well, where is it? A cave, of course?"

"No, excellency; and you, effendi," he continued, turning to the professor, "will be delighted."

"What do you mean, my good fellow?" said the professor warmly.

"That you will have your wish. There is no other place likely, and it seems to me that this band of men have made the old ruined stronghold their lurking-place, and you will see the ruins after all."

"What?" cried Mr. Preston excitedly.

"I am not sure, excellency, for they may be only going to pass them on our way elsewhere; but we are now journeying straight for the grand old remains we sought."

"Then, I don't care what ransom I have to pay," said the professor eagerly. "Lawrence, my dear boy— Burne—this is not a misfortune, but a great slice of luck."

"Oh! indeed! is it?" said the old lawyer sarcastically. "I should not have known."

It proved to be as Yussuf had anticipated, for, just as the sun was sinking below the mountains, the shelf of a path was continued along by the brink of a terrible precipice which looked black beneath their feet, and after many devious windings, it ended as it were before a huge pile of limestone, at the foot of which rocks were piled up as if they had suddenly been dashed down from some tremendous tremor of the mountains.

"Where are we going?" said the professor.

"Up to the top of that great pile," said Yussuf.

"But are the ruins there?"

"Yes, effendi."

"And how are we to get there?"

"You will see, excellency. It is quite right. This is the robbers' home, where they could set an army at defiance."

"But we can't get up there," said Lawrence, gazing at the dizzy height.

As he spoke, the foremost horseman seemed to disappear, but only to come into sight again, and then it became evident that there was a zigzag and winding path right up to the top of the huge mass of rock which towered up almost perpendicularly in places, and, ten minutes later, Lawrence was riding up a path with so awful a precipice on his right that he closed his eyes.

But the next minute the fascination to gaze down was too strong to be resisted, and he found himself looking round and about him, almost stunned by the aspect of the place. But the sure-footed Turkish ponies went steadily on higher and higher round curves and sharply turning angles and elbows, till at last at a dizzy height the foremost horseman rode in between two masses of rock surmounted by ruined buildings. Then on across a hideous gap of several hundred feet deep, a mere split in the rock bridged with the trunks of pine-trees, but awful to contemplate, and making the travellers hold their breath till they were across, and amid the gigantic ruins of an ancient stronghold.

"Stupendous!" cried the professor, as they rode on amidst the traces of the former grandeur of the place.

"How bitterly cold!" said the professor.

"We are to dismount here," said Yussuf quietly, "and go into this old building."

They obeyed, glad to descend from their horses, which were taken away, and then they were ushered to a great stone-built hall where a fire was burning, which seemed cheery and comfortable after their long ride.

There were rugs on the floor, the roof was sound, and the window was covered by a screen of straw which made the place dark save for the warm glow of the fire, near which a little Turkish-looking man was seated, and a largely proportioned Turkish woman reclined on a rough kind of divan.

"These are to be our quarters, effendi," said Yussuf, after a brief colloquy with the chief, who had accompanied them, "and these are our fellow-prisoners. But he warns me that if we attempt to escape we shall be shot, for there are sentries on the watch."

"All right," said Mr. Burne approaching the fire; "tell him not to bother us to-night, only to give us the best they've got to eat, or else to let us have our baggage in and leave us to shift for ourselves."

Just then an exclamation escaped the big Turkish woman, who sprang to her feet, and ran and caught the professor's hand.

"Mr. Preston!" she cried. "Do you not know me?"

"Mrs. Chumley!" cried the professor. "You here!"

"Yes, we've been prisoners here for a month. Charley, you lazy fellow, get up; these are friends."

"Oh, are they?" said the little Turk, rising slowly. "Well, I'm jolly glad of it, for I'm sick of being here. Hallo, young Lawrence, I've often thought about you; how are you? Getting better? That's right. See you are. How do, Preston? How do, Mr. Burne? I say! Ha-ha-ha! You're all in for it now."

"For shame, Charley, to talk like that," cried the lady. "Come up to the fire all of you. I am very glad to see you here."

"Oh, you are, eh, madam?" said the old lawyer sharply, as he warmed his hands over the blaze.

"Well, I do not mean that," said the lady; "but it is always pleasant to meet English people when you are far from home."

Just then the robber chief nodded, said a few words to Yussuf, and the prisoners were left alone.

CHAPTER XXXVII.

A QUESTION OF RANSOM.

ICE state of affairs this, Mr. Preston," said the little prisoner holding out his arms. "Here's a dress for a gentleman;" and he displayed the rags of Turkish costume he wore. "Chaps saw me at my club now."

"Charley, will you hold your tongue," cried his lady angrily. "How am I to explain our position if you will keep on chattering so?"

"But, my darling—"

"Will you be quiet, Charley. Look here, Mr. Preston," she continued, "it's just three weeks ago, as we were travelling in this horrible country at least ten miles away, we were seized by these horrid men, and brought here. They've taken everything we had, and given us these miserable clothes, and every night they come to us and say—"

"They'll cut off our heads to-morrow morning."

"Will you be quiet, Charley," cried the lady, stamping her foot. "How am I to explain? Am I not always telling you what a chatter-box you are."

"Yes, my dear, always."

"Silence, sir! Mr. Preston," she continued, as her little husband went softly to Lawrence, and drew him aside to go on whispering in his ear—" Mr. Preston, no one knows what we have suffered. As I was saying—I hope you are listening, Mr.—Mr.—Mr.—Mr.—"

"Burne, ma'am," said the old lawyer bowing.

"Oh, yes, I had forgotten. Mr. Burne. I beg your pardon. As I was saying they come every night, and say that to-morrow morning they will cut off our heads and send them to Smyrna as an example, if our ransom does not come."

"Your ransom, madam?" said the professor.

"Yes. Five thousand pounds—three for me and two for poor Charley; and though we have sent for the money, it does not come. Isn't it a shame?"

"Scandalous, madam."

"And you can't tell how glad I am to see you here. Have you brought the money?"

"Brought the money, ma'am? Why, we are prisoners too."

"Oh, dear me, how tiresome!" cried the lady. "I thought you were at first; and then I thought you were sent with our ransom. What are we to do? Mr. Burne," she continued, turning to him, "you said you were a lawyer. Pray, send for these people at once, and tell them that they will be very severely punished if they do not set us at liberty."

"My dear madam," said the old lawyer, "I am only just getting myself thawed, and I have had nothing but snuff since breakfast. I must have some food before I can speak or even think."

Meanwhile little Mr. Chumley was whispering to Lawrence on the other side of the fire, and relating all

his troubles. "Taken everything away, sir," he said
—"watch, purse, cigars, and I actually saw the scoundrel who is at the head of them smoking my beautiful
partagas that I brought with me from England. I
say, what had we better do?"

"Try and escape, I suppose," said Lawrence.

"Escape! Look here, young man; are you a fly, or
a bird, or a black beetle?" whispered the little man.

"I think not," said Lawrence laughing.

"Then you can't get away from here, so don't think
it. Why, it's impossible."

Just then the fierce-looking chief entered, followed
by a man carrying a great smoking dish, and as the
leader drew near the fire, Lawrence bit his lip, for he
saw that the tall ruffian was wearing his sword, and
carrying his handsome gun in the hollow of his arm.

The chief turned to Yussuf, who was seated in one
corner of the room, and said a few words to him.

Yussuf rose and addressed his little party in a low
voice.

"The brigand captain says, excellencies, that you
are to be prepared to send in one of his men to-morrow
morning as messenger to your agent where you like.
You are to write and say that, if injury is done to the
messenger, you will be killed. The messenger is to
bring back six thousand pounds—two for each of you—
as a ransom, and that, upon the money being paid, you
will be set free."

"And if the money be not paid, Yussuf, what then?"
said the professor quietly.

"The chief says no more, excellency."

"But he will to-morrow or next day," cried Mr.
Chumley. "He'll say that if the money is not paid
he'll—"

" Will you be quiet, Charley?" cried his wife. "How you do chatter, to be sure! Are you going to send for the money?"

" I don't know yet," said the professor smiling. " I must think over our position first."

" But, Mr. Burne!" cried the lady.

" My dear madam," said Mr. Burne, " I can say nothing till after supper. Here is a dish of fowl and rice to be discussed before we do anything else. Here, Snooks, Brown, Hassan, Elecampane—what's your name?—lay the cloth and bring some knives and forks."

The man addressed did not stir. He had placed the smoking brass dish upon a stone near the fire, and with that his duties seemed to be ended.

" They won't give you any knives or forks," said little Mr. Chumley.

"Will you be quiet, Charley?" cried his lady. " No, gentlemen, you will have to sit down all round the dish like this, and eat with your fingers like pigs."

" Pigs haven't got any fingers," whispered little Chumley to Lawrence. " Come along."

" What is he whispering to you, Master Lawrence?" said the lady sharply. " Don't take any notice of what he says. He talks too much and thinks too little. If he had thought more and said less we should not be in this predicament."

The chief and his follower had passed silently behind the great rug stretched over the doorway, and, led by their hunger, the prisoners all sat down round the dish "like this," to use Mrs. Chumley's words—*this* being tailor fashion, or cross-legged à la Turcque; and then, in very primitive fashion, the supper of poor stringy fowl and ill-cooked rice began.

The food was very poor, the bread being heavy and

black; but all were too hungry to be particular, and at last the dish was completely finished, and conversation respecting their position began, while Yussuf sat aside and waited patiently to be questioned.

"Look here, Yussuf," said the professor at last; "what is to be done?"

"I fear, excellency," replied the guide, "that the only way of escape is by paying the ransom."

"But, man, it is ruinous, and they dare not injure us. Why, if the English people knew of our position, troops would be sent to our assistance."

"And the brigands would resent their coming by killing you and your friends, excellency."

"They would not dare, Yussuf."

"I'm afraid they would, effendi. They are utterly reckless scoundrels, the sweepings of the country, and they are so powerful, and secure here that they laugh at the law, such law as we have in this unhappy land."

"But such a state of affairs is monstrous, sir," said Mr. Burne. "I am a lawyer, sir, and I ought to know."

"It is monstrous, excellency," said Yussuf; "but these men are outlaws. You see what a stronghold they have if it came to a fight; but your friends or the government would not dare to let it come to a fight, for if they did they would be slaying you."

"Tchah!" cried Mr. Burne; "this is about the knottiest case I ever did meet. I say, you, Lawrence, a nice position you have placed us all in."

"I, Mr. Burne!" cried the lad wonderingly.

"Yes, sir, you. If you had only been quite well, like a reasonable boy of your age, we should not have come out here, and if we hadn't come out here we should not have been in this mess. There, I'm too tired to talk. Good-night."

He threw himself down upon one of the rugs and was asleep directly, while the professor walked to the doorway, and found two fierce-looking sentries outside, one of whom menacingly bade him go back.

He spoke in the Turkish language; but his manner made his meaning plain, so Mr. Preston went back to the fireside, and sat talking to the Chumleys and Lawrence till the latter fell fast asleep; and at last, in spite of the peril of his position, the professor grew so weary that the account of the Chumleys' troubles began to sound soothing, and, what with the long day's work, the exposure to the keen mountain air, and the warmth of the fire, he too fell asleep, and silence reigned in the ancient structure that had been made their prison.

CHAPTER XXXVIII.

SUGGESTIONS OF ESCAPE.

THE morning broke so bright and clear, and from the window there were so many wonders of architecture visible in the old stronghold, that the professor and Lawrence forgot for the time that they were prisoners, and stood gazing out at the wonderful scene.

Where they had been placed was evidently a portion of an old castle, and looking down there were traces of huge buildings of the most solid construction, such as seemed to date back a couple of thousand years, and yet to be in parts as strong as on the day they were placed and cemented stone upon stone.

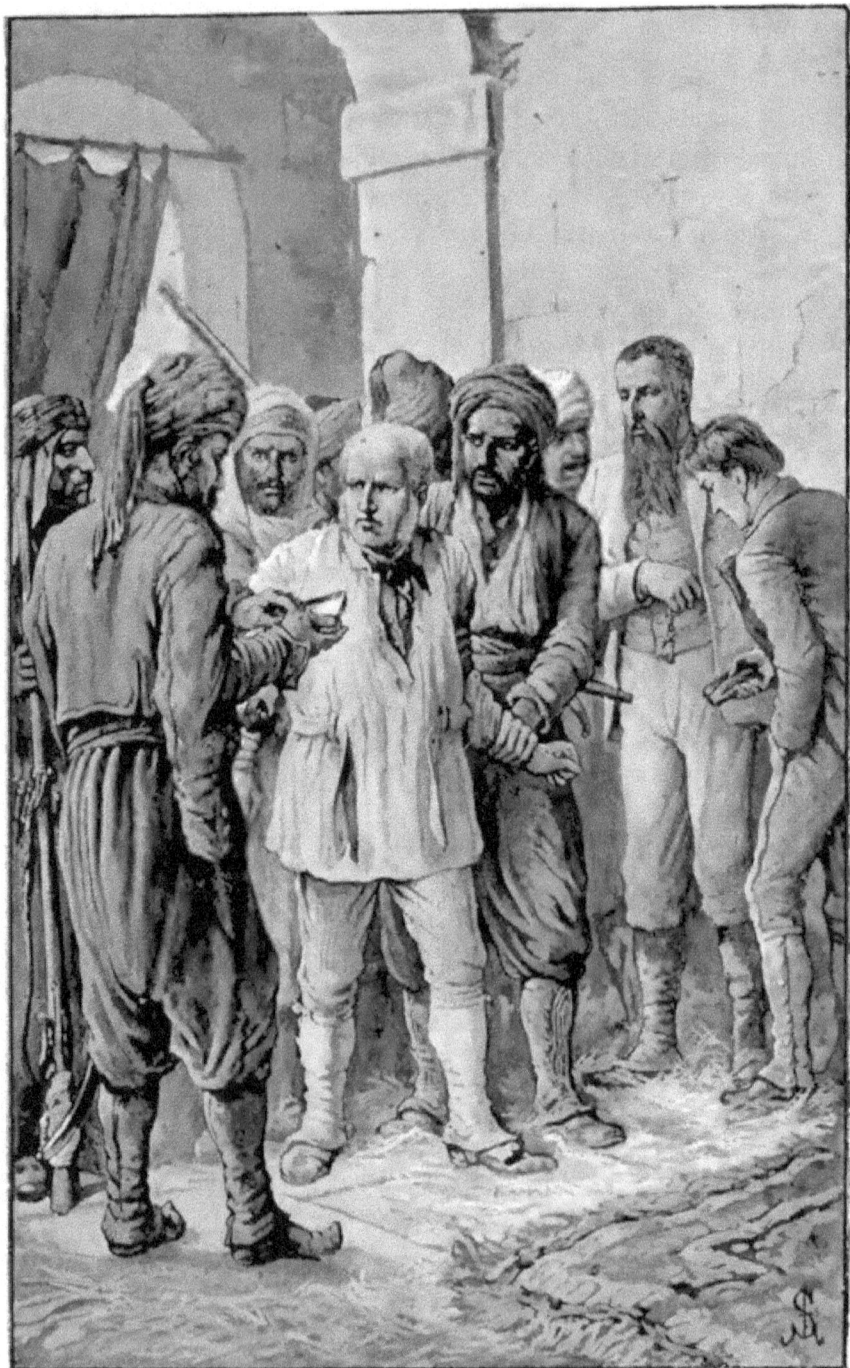

THE BANDIT CHIEF INVESTIGATES MR. BURNE'S SNUFF-BOX.

Huge wall, tremendous battlement, and pillared remains of palace or hall were on every side, and as they gazed, it seemed to them that they could easily imagine the presence of the helmeted, armoured warriors who had once owned the land.

The sun was so glorious that the professor proposed a look round before breakfast.

" Never mind the inconvenience, Lawrence," he said, " we have fallen into a wonderful nest of antiquities, worth all our journey and trouble. Here, come along."

They went to the doorway, drew the great rug hanging before it aside, and were stepping out when a couple of guns were presented at their breasts, and they were angrily bidden to go back.

It was a rude reminder that they were no longer upon a touring journey, and the fact was farther impressed upon them, after a breakfast of yaourt or curd, bread, and some very bad coffee, by a visit from the chief and half a dozen men.

Yussuf was called upon to interpret, and that which he had to say was unpalatable enough, for he had to bid them empty their pockets, and pass everything they possessed over to their captors.

Watches, purses, pocket-books, all had to go; but it was in vain to resist, and everything was handed over without a word, till it came to Mr. Burne's gold snuff-box, and this he slipped back into his pocket.

The attempt to save it was in vain; two sturdy scoundrels seized him, one on each side, and the snuff-box was snatched away by the chief himself.

He uttered a few guttural sounds as he opened the box, and seemed disappointed as he found therein only a little fine brown dust, into which he thrust his finger and thumb.

He looked puzzled and held it to his nose, giving a good sniff, with the result that he inhaled sufficient of the fine dust to make him sneeze violently, and scatter the remainder of the snuff upon the earth.

Mr. Burne made a start forward, but he was roughly held back, and the chief then turned to Yussuf.

" Tell them," he said in his own tongue, "to write to their friends, and ask for the ransom—two thousand pounds each, and to say that if the money is not given their heads will be sent. Bid them write."

The fierce-looking scoundrel turned and stalked out of the place with his booty, and the moment he was free, Mr. Burne dropped upon his knees and began sweeping the fallen snuff together in company with a great deal of dust and barley chaff, carefully placing the whole in his handkerchief ready for clearing as well as he could at his leisure.

" That's just how they served us," said Mrs. Chumley dolefully. " I thought they would treat you the same."

" So did I," said her husband dolefully. " They've got my gold repeater, and—"

" Now, Charley, don't—don't—don't bother Mr. Preston about that miserable watch of yours, and I do wish you wouldn't talk so much."

" But we must talk, madam," cried Mr. Burne. "Here, you, Yussuf, what's to be done?"

" I can only give one piece of advice, effendi," said Yussuf gravely; " Write."

" What, and ruin ourselves?"

" Better that than lose your life, effendi," replied the guide. " These people are fierce, and half savage. They believe that you have money, and they will keep their word if it is not sent."

"What, and kill us, Yussuf?" said Lawrence, with a horrified look.

"Not if I can save you, Lawrence effendi," said Yussuf eagerly. "But the letters must be sent. It will make the villains think that we are content to wait, and put them off their guard. Preston effendi, it is a terrible increase of the risk, but you will take the lady?"

"Take the lady?"

"Hush! When we escape. Do not say more now; we may be overheard. Write your letters."

"Then you mean to try and escape."

"Try and escape, effendi?" said Yussuf with a curious laugh; "why, of course."

"What will you do?"

"Wait, excellency, and see. There are walls here, and I think places where we might get down past the guards with ropes."

"And the ropes?"

Yussuf laughed softly, and stared at the rugs as he said quietly:

"I can see the place full of ropes, your excellency; only be patient, and we'll try what can be done in the darkness. Write your letters now."

Mr. Preston had to appeal to the sentries, through Yussuf, for the necessary writing materials, and after a good deal of trouble his own writing-case, which had been in the plundered baggage, was brought to him. He wrote to the vice-consul, Mr. Thompson, at Smyrna, telling of their state, and asking advice and assistance, telling him, too, how to obtain the money required if diplomacy failed, and the ransom could not be reduced.

This done, and a similar letter being written by Mr.

Burne, the sentry was again communicated with, and the despatches sent to the chief.

An hour later there was a little bustle in the open space before their prison, and a couple of well-armed men mounted their horses, the chief standing talking to them for a few minutes, as if giving them final instructions.

He then summoned his prisoners, and spoke to Yussuf, bidding him ask Mr. Burne, whose wonderful head-dress won for him the distinction of being considered the most important personage present, whether he would like to make any addition to his despatch; for, said he:

"I have told the people that any attempt at rescue means your instant death. I will wait any reasonable time for your ransoms, and you shall be well treated; but I warn you that attempts to escape will be death to you. That is all."

"Wait a minute, Yussuf," said Mr. Burne. "Tell him he can keep the snuff-box and welcome, but he has a canister of best snuff in the package that was on the brown pony. Ask him to let me have that."

"Yes," said the chief, on hearing the request, "it is of no use to anyone. He can have it. What a dog of a Christian to take his tobacco like that! Anything else?"

"Yes," said Mr. Preston, on hearing the reply, "tell him to send his men to watch me as much as he likes, but I want leave to inspect the old ruins and to make drawings. Tell him I will not attempt to escape."

"No, effendi," said Yussuf, "I will not tell him that, but I will ask the first;" and he made the request.

"What! is he—one of the idiot giaours who waste their time in seeing old stones and imitate them upon paper?"

"Yes, a harmless creature enough," said Yussuf.

"So I suppose, or he would have fought. Well, yes, he can go about, but tell him that if he attempts to leave my men behind they will shoot him. Not that he can get away, unless he has a djin to help him, or can fly," he added with a laugh.

He walked to his men, gave them some further instructions, and they saw the two ambassadors go in and out among the ruins till they passed between two immense buttresses of rock, and then disappear down the perilous zigzag path that led to the shelf-like way.

"Yes," said Yussuf, looking at Mr. Preston, and interpreting his thoughts, "that is the only way out, excellency, but I do not despair of making our escape. It must be a long time before arrangements can be made for your release, and the winter comes early here in these high places."

"Winter?" cried Lawrence.

"Yes," said Yussuf. "It is fine and sunny one day, the next the snow has fallen, and a place like this may be shut off from the plains below for months. You do not wish to pass the winter here, Lawrence effendi?"

"I don't think I should mind," replied the lad, "everything is so fresh, and there is so much to see."

"Well, now they are giving me leave to go about," said Mr. Preston thoughtfully, "I think I could spend some months in drawing and writing an account of this old city, especially if they would let me make some excavations."

"But his excellency, Mr. Burne?" said Yussuf.

"Oh! I've got my snuff—at least I am to have it, and if they will feed us well I don't suppose I should mind very much. The fact is, Preston, I've been working so hard all my life that I like this change. Doing nothing is very pleasant when you are tired."

"Of course it is," said the professor smiling.

"And so long as there's no nonsense about cutting off men's heads, or any of that rubbish, I rather like being taken a prisoner by brigands. I wonder what a London policeman would think of such a state of affairs."

"My masters are submitting wisely to their fate," said Yussuf gravely; "and while we are waiting, and those people think we are quite patient, I shall come with his excellency Preston, and while he draws I shall make plans, not of the city, but how to escape."

Further conversation was cut short by the coming of Mr. and Mrs. Chumley, who eagerly asked—at least Mr. Chumley wished to ask eagerly, but he was stopped by his lady, who retained the right—what arrangements had been made. And she was told.

"Oh, dear!" she sighed, "then that means weary waiting again. Oh, Charley! why would you insist upon coming to this wretched land?"

Mr. Chumley opened his mouth in astonishment, but he did not speak then, he only waited a few minutes, and then took Lawrence's arm, and sat whispering to him apart, telling him how Mrs. Chumley had insisted upon coming to Turkey when he wanted to go to Paris, and nowhere else, and that he was the most miserable man in the world.

Lawrence heard him in silence, and as he sat he wondered how it was the most miserable man in the world could look so round and happy and grow so fat.

CHAPTER XXXIX.

YUSSUF HAS HIS WITS ABOUT HIM.

HE weather was cold up there in the mountains, and it froze at night; but the sun was hot in the daytime, and the sky was mostly of a most delicious blue. The chief always seemed to be scowling, watchful, and suspicious, but the prisoners had nothing but their captivity to complain about. Rugs in abundance—every one of them stolen—were supplied for bedding and keeping out the cold night air that would have penetrated by door or window. Upon proper representations being made by Yussuf the food supply was better, the guide installing himself at once as cook, to Mr. Chumley's great delight; and agreeable dishes—pilaf, curry, kabobs, and the like—were prepared, with excellent coffee and good bread, while the scowling sentries became more agreeable, and took willingly to their duties, on finding that satisfactory snacks were handed to them, and hot cups of coffee on the bitter nights when they sat watching in their sheepskin or goatskin cloaks.

As for the professor, in two days he had forgotten that he was a prisoner, and Lawrence was the best of friends with the evil-looking guards, who followed them with loaded guns to some old ruinous patch of wall, fortification, or hall. Here the professor was in his element, drawing, planning, and measuring, longing the while to set a dozen strong-armed men to work digging up the stones embedded in the earth—

a task which he was sure would be rewarded by the
discovery of many objects of antiquity.

Parties of the brigands went out now and then, but
it was evident that their object was merely to forage,
large quantities of barley being brought in, and some
of the old buildings being utilized for stores.

These seemed to be well supplied, and the com-
munity was preparing for the coming winter, so Yussuf
told Lawrence—for the days when no food would be
obtainable perhaps for months.

Everyone seemed to lead a careless nonchalant life,
the prisoners they had taken would, no doubt it was
considered, bring in sufficient to make this a prosperous
year's work, and till the ransoms were paid there was
little more to do.

The days glided by, and the watch over the prisoners grew less rigid. There was apparently only
one way out of the stronghold, and that was always
carefully guarded; and as it was evident to the captors that the professor and his companions were bent
upon studying the place, the guards used to sit down
upon some heap of old stones, with their guns across
their knees, and smoke and sleep, while drawings were
made, and inscriptions copied.

Yussuf became quite a favourite, for he was a cook,
and often showed the brigands' wives how to make
some savoury dish; but for the most part he was busy
helping the professor, carrying his paper, cleaning
stones, or performing some such office.

And so the days glided by, with the professor perfectly contented, the old lawyer apparently little
troubled so long as his snuff held out, and Lawrence
growing sturdier, and enjoying the feeling of health
more and more.

The only discontented people were the Chumleys, the gentleman complaining bitterly about the absence of news, and the lady because her husband would chatter so incessantly.

"I say, Yussuf," said Lawrence one night as he sat talking to the guide, "they won't cut off our heads, will they?"

Yussuf shook his head.

"I have only one dread," he replied; "and that is of an attempt being made to rescue us."

"I don't see anything to be afraid of there," said Lawrence laughing.

"But I do," said the Turk seriously. "If an attack were made, those people would become fierce like dogs or rats at bay, and then they might take our lives."

"They would not without, then?"

"No," said Yussuf; "they would threaten, and hold out for a heavy ransom, but if the friends that have been written to are clever, they will make the ransom small, and we shall be freed. But it may take a long time, for the brigands will hold out as long as they think there is a chance of getting a large sum. They are safe here; they have abundant stores, and nothing to do: they can afford to wait."

"Well, I'm sure Mr. Preston is in no hurry," said Lawrence; "nobody is but the Chumleys."

"And I," said Yussuf smiling.

"You? why, I thought you were happy enough. You haven't said a word lately about escaping."

"No," replied Yussuf smiling; "but sometimes those who are so quiet do a great deal. I am afraid of the winter coming with its snow and shutting us in for months when we could not escape, for, even if the

snow would let us pass, we should perish in the cold.
I have been hard at work."

"You have, Yussuf? What have you been doing?
Oh, I know; making plans."

"And ropes," said Yussuf gravely.

"Ropes? I have seen you make no ropes."

"No, because you were asleep. Wait a moment."

He rose quietly and walked to the entrance, drawing the rug that hung there aside and peering out, to
come back as softly as he left his seat, and glancing at where the professor, wearied out with a hard
day's work, was, like his companions by the fire, fast
asleep.

"The guards are smoking out there, and are safe,"
said Yussuf. "See here, Lawrence effendi, but do not
say a word to a soul."

"I shall not speak," said Lawrence.

Yussuf gave another glance at the Chumleys, and
then stepped to a corner of the great hall-like place
which formed their prison, drew aside a rug on the
floor, lifted a slab of stone, and pointed to a coil of
worsted rope as thick as a good walking-stick, and
evidently of great length.

It was only a few moments' glance, and then the
stone was lowered, the dust swept over it, and the rug
drawn across again.

"You see I am getting ready," said Yussuf.

"But what are we going to do?"

"I have been watching and waiting," whispered the
guide, "and I have found a place where we can descend
from the old wall over the great defile."

"But it is so awful a place, Yussuf."

"Yes, it is awful; but there is a ledge we can reach,
and then creep along and get beyond the sentries.

Then all will be easy, for we can get a long way some dark night before the alarm is given, and in the day we can hide. Of course we must load ourselves with the food we have saved up."

"Yes, yes, of course," said Lawrence thoughtfully; "but Mrs. Chumley, she would not go down a rope."

"Why, not?" said Yussuf quietly; "she talks like a man."

"When are you going to try, then?" said Lawrence excitedly.

"In about ten days. I shall be ready then, and the nights will be dark. But, patience—you must not be excited."

"But you will tell Mr. Preston?"

"Yes; to-morrow night, when I have finished my first rope. Go to sleep now."

"And you, Yussuf?"

"Oh, I am going to work," he said smiling. "See, my material is here."

He drew out a handful of worsted threads which were evidently part of a rug which he had unravelled, and as soon as Lawrence had lain down, the Turk walked to the darkest corner of the building, and Lawrence could just make out that he was busy over something, but he was perfectly silent.

CHAPTER XL.

A GRAND DISCOVERY.

I' was the very next day that the professor took his paper, rule, and pencils down to a building that seemed to have been a temple. It was at the very edge of the tremendous precipice, and must once have been of noble aspect, for it was adorned with a grand entrance, with handsomely carved columns supporting the nearly perfect roof, and the wonder was that the brigands had not utilized it for a dwelling or store. But there it was, empty, and the professor gazed around it with rapture.

The guards stood at the entrance leaning against the wall watching him and Lawrence carelessly, and then, going out into the sunshine, they picked out a sheltered spot, and sat down to smoke.

The professor began to draw. Soon afterwards Mr. Burne sat down on a broken column taking snuff at intervals, and Yussuf seated himself with his back to the doorway, drew some worsted from his breast, and began to plait it rapidly, while Lawrence went on investigating the inmost recesses of the place.

"Come and look here, Yussuf," he cried at the end of a few minutes, and the Turk followed him to a part of the building behind where an altar must have stood and pointed down.

"Look here," he said; "this stone is loose, and goes down when I stand upon that corner. It's hollow, too, underneath."

He stamped as he spoke, and there was a strange echoing sound came up.

"Hush!" said Yussuf quickly, and he glanced round to see if they were observed; but they were hidden from the other occupants of the place; and, stooping down, Yussuf brushed away some rubbish, placed his hands under one side of the stone where it was loose, and lifted the slab partly up.

The air came up cool and sweet, so that it did not seem to be a vault; but it was evidently something of the kind, and not a well, for there was a flight of stone steps leading down into the darkness.

It was but a moment's glance before Yussuf lowered the stone again, and hastily kicked some rubbish over it, and lowered a piece of an old figure across it so as to hide it more.

"What is it?" said Lawrence quickly.

"I do not know," replied Yussuf. "It is our discovery. It may be treasure; it may be anything. Say no word to a soul, and you and I will get a lamp, escape from the prison to-night, and come and examine it, and see what it is. It may be a way out."

Lawrence would gladly have gone on at once, but Yussuf signed to him to be silent; and it was as well, for he had hardly time to throw himself down on a block of stone, and sham sleep, when the guards came sauntering in and looked suspiciously round. Then, not seeing two of their prisoners, they came on cautiously, and peered over the stones that hid them from where the professor was drawing, to find Yussuf apparently asleep, and Lawrence sharpening his pocket-knife upon a stone.

One of the men came forward and snatched the knife away, saying in his own tongue that boys had no business with knives, after which he stalked off and returned to his old place outside.

"You see," said Yussuf quietly, "it was no time now for examining the place; wait till night."

For the first time since he had been a prisoner the hours passed slowly to Lawrence. It seemed as if it would never be night, and every time he met the professor's or Mr. Burne's eye, they seemed to be taking him to task for keeping a secret from them.

Then, too, Mrs. Chumley appeared to be suspecting him, and Chumley drew him aside as if to cross-examine him: but it was only to confide a long story about how severely he had been snubbed that day for wanting to follow the professor to the ruins where he was making his drawings.

At last, though, the guards had thrust in their villainous faces for the last time, according to their custom, and all had lain down as if to sleep.

An hour must have passed, and Lawrence lay with his heart beating, waiting for a summons from Yussuf; but it seemed as if one would never come, and the lad was about to give up and conclude that their guide had decided not to go that night, when a hand came out of the darkness and touched his face, while a pair of lips almost swept his ear, and a voice whispered:

"Rise softly, and follow me."

Lawrence needed no second invitation, and, rising quickly, he followed Yussuf to where the rug hung over the door.

"Bend down low, and follow me," whispered the Turk. "The guards are nearly asleep."

He drew the rug a little on one side, and Lawrence saw where the two men were huddled up in their sheep-skin cloaks.

"Do as I do," whispered Yussuf.

The moon was shining, and the part where the guards sat was well in the light; but a black shadow was cast beneath the walls of the great building, and by stooping down and keeping in this, the evading pair were able to get beyond the ken of the guards, and though lights shone out from one ruined building, whether from fire or lamp could not be told, not a soul was about, and they were able to keep on till the inhabited part was left behind and the old temple reached.

"It was a dangerous thing to do, Lawrence effendi," said the guide. "I repented promising to bring you, for the men might have fired."

"Never mind that," whispered Lawrence. "We are safe now. Have you brought a light?"

"Yes," was the reply; and, by the moonlight which shone through a gap, Yussuf led the way among the broken stones to the back of the old altar, where, after feeling about, he found the side of the stone, lifted it right up, and leaned it against a broken column.

Then, after a word of warning, he stooped down and struck a match, but the draught that blew up the opening extinguished it on the instant.

Another and another shared the same fate, after giving them a glimpse of a ragged set of stone steps; and as it was evident that no light could be obtained that way, Yussuf took the little lamp he had brought into a corner of the building, lit it, and sheltering it inside his loose garment, he came back to where Lawrence waited listening.

"I'll go first," said Yussuf. "Mind how you come."

He lowered himself into the hole, and descended a few steps.

"It is quite safe," he said. "Come down;" and Lawrence descended to stand by his side.

"Shelter this lamp a minute," whispered Yussuf. "I must close the stone, or the light will be out."

Lawrence took the lamp, the perspiration standing on his forehead the while, as he felt that this was something like being Aladdin, and descending into the cave in search of the wonderful lamp.

"Suppose," he thought, "that Yussuf should step out and leave him in this horrible place to starve and die. Nobody would ever guess that he was there, and no one would hear his cries. What was the place—a tomb? And had Yussuf gone and left him?"

There was a low dull hollow sound as the stone descended into its place, and a cry rose to the lad's lips, but it had no utterance, for Yussuf said softly from above:

"Now you may show the light, and we can see where we are."

Lawrence drew a breath of relief as he took the light from his breast, and saw that he was standing upon a very rough flight of stone steps, with the rugged wall of rock on either side.

Yussuf took the lamp and held it up, showing a rough arch of great stones over their heads, and the square opening over a rough landing where they had descended, while on either side the rock looked as if at some time it had been split, and left a space varying from four to six feet wide, the two sides being such that, if by some convulsion of nature they were closed, they would have fitted one into the other.

"Follow close behind me," said Yussuf. "This must lead into some vault or perhaps burial-place. You are not frightened?"

"Yes, I am," said Lawrence in a low tone.

"Shall we go back?"

"No, but I cannot help being a little alarmed."

Yussuf laughed softly.

"No wonder," he said. "I feel a little strange myself. But listen, Lawrence; what we have to fear is a hole or crack in the rock into which we might fall, so keep your eyes on the ground."

But their path proved very easy, always a steep descent, sometimes cut into stairs, sometimes merely a rugged slope, and always arched over by big uncemented stones.

No vault came in sight, no passage broke off to right or left; it was always the same steep descent—a way to some particular place made by the ancients, who had utilized the crevice or split in the rock, and arched it over to make this rugged passage.

"I think I understand," said Yussuf, when they had gone on descending for quite three hundred yards.

"What is it?" said Lawrence; "a tomb?"

"No."

"A treasure chamber?"

"No."

"What, then?"

"There must be a spring of good water somewhere down at the bottom, and this was of great value to the people who built this place on the rock. Shall we go any farther?"

"Yes, I want to see the spring," said Lawrence. "I am not so frightened now."

"There is quite a current of air here," said Yussuf, when they had descended another hundred yards or so. "The spring must be in the open air, and out by the mountain side."

Lawrence was too intent upon his feet to answer, and they descended another fifty yards, when Yussuf stopped, for the way was impeded by a piled-up mass of fallen stones, and on looking up to see if they were from the roof they found that the arching had ceased, and that the roof was the natural rock of wedged-in masses fallen from above.

"We can get no farther," said Yussuf, holding the lamp above his head.

"Look, look!" said Lawrence softly; "there is a light out there."

Yussuf looked straight before him; and placing the lamp upon the ground, and shading it with his coat, there, sure enough, not more than a dozen yards away, was a patch of light—bright moonlight.

"I was wrong," said Yussuf calmly; "this is not the way to a spring, but a road from that temple down to some pathway along by the side of the mountain, and closed up by these fallen stones. Lawrence effendi, we shall not want my ropes to descend from the walls. You have found a way out of the old place that has lain hidden for hundreds of years."

"Do you think so?"

"Yes: and that we have only to set to work and clear away these stones sufficiently to reach the entrance, and then we can escape."

"Let us begin, then, at once," cried Lawrence joyously.

"No; we will go back now, and examine the way, so as to make sure that our course up and down is safe. Then we will get back, and be satisfied with our night's work."

"Yes," said Yussuf, when he reached the stone again; "it is all quite plain. I could come up and down here in the dark, and there will be light enough at the bottom in the daytime to see what to do."

He raised the stone after extinguishing his lamp, and they both stepped out; the stone was lowered into its place, a little earth and dust thrown over it and a few fragments of rubbish, and then the midnight wanderers stole back to the prison, but only to stop short in the shadow with Lawrence chilled by horror. For, as they were about to step up to the portal, one of the guards yawned loudly, rose, and walked to the rug, drew it aside, and looked in.

He stood there gazing in so long, that it seemed as if he must have discovered that there were absentees; but, just as Lawrence was in despair, he dropped the curtain, walked back to his companion, and sat down with his back to the portal.

Yussuf wasted no time, but glided along in the shadow, and Lawrence followed; but as he reached the portal he kicked against a piece of loose stone and the guards sprang up.

Lawrence would have stood there petrified, but Yussuf dragged him in, hurried him across the interior, threw him down, and took his place behind him.

"Pretend to be asleep," he whispered; and he turned his face away, as the steps of the guards were heard, and they lifted the rug curtain and came in with a primitive kind of lantern, to look round and see if all were there, being satisfied on finding them apparently asleep, and going back evidently believing it was a false alarm.

"Safe this time, Yussuf," whispered Lawrence.

"Yes," said the guide. "Now sleep in peace, for you have discovered a way to escape."

CHAPTER XLI.

THE TIME FOR FLIGHT.

ND you are sure, Yussuf?" said Mr. Preston two days later.

"Yes, effendi. I have been there alone twice since, and in a few hours I had moved enough stones to let me through to the light, and in a few hours more I can make the passage so easy that a lady can go through."

"And where the light shines in?"

"Is just over a narrow rugged path leading down the mountain — a way that has been forgotten. Effendi, after I have been there once again the way is open, and though the path is dangerous it will lead to safety, and we must escape."

"When?" said Mr. Preston eagerly.

"As soon as we can collect a little food—not much, but enough to carry us to the nearest village where we can get help."

"And our goods—our property?"

"Must stay, excellency. Once you are all safe we can send the soldiery by the path by which we left, for the brigands will not know how we have escaped."

"Well, I can save my drawings," said the professor, "and they will be worth all the journey, as we have no ransom to pay."

The next day Mr. Burne was let into the secret, but it was decided not to tell the Chumleys till they were awakened on the night of the attempt.

It was hard work to keep down the feeling of elation so as not to let the chief see that the captives were full of hope, for he came day by day to visit them and complain about the length of time his messengers were gone.

But the secret was well kept, and those who shared it, in obedience to Yussuf's suggestion, began to store away portions of their provisions so as to be prepared at any moment for a journey which might take them for many days through the mountains away from village or beaten track.

" I shall leave this place with regret," the professor said with a sigh; " but I must say I do not relish paying for my stay with every shilling I have scraped together during my life."

" No. Let's get away, Preston," said Mr. Burne. " Oh, if I could only commence an action against these scoundrels for our imprisonment! I'd make them smart."

They were sitting together among the ruins, and their thoughts naturally reverted to Yussuf and his reticent ways, for two days had passed since he had made any communication, and he had seemed to be more retiring than ever.

The sun was shining brightly, and warmed the stones where they sat, but the air seemed to be piercingly cold, and Mr. Burne shivered more than once, and got up to walk about.

" I shall not be sorry to get down out of the mountains," he said. " What do you say, Lawrence?"

" Oh! I've liked the stay up here very well, it has all been so new and different; and besides, I have been so well, and I feel so strong."

" Yes, you are better, my boy," said Mr. Burne, nodding his head approvingly.

" I used to feel tired directly I moved," continued

Lawrence, "but now I scarcely ever feel tired till quite night. Yussuf says it is the mountain air."

"Yes," said the professor dryly, "it is the mountain air. Where is Yussuf?"

"Here, excellency," said their guide; and they all started with surprise, he had approached so quietly. "I was coming to tell you that I have been up to the top of the old temple, and have at length traced the ancient path. I have only seen parts of it here and there, but I can make out the direction it takes, and it is right opposite to that by which we came."

"But where does it lead?" said the professor.

"Away west, effendi—where, I cannot say; but let us get out of this place and I will lead you in safety somewhere."

"But the old path — is it very dangerous?" said Mr. Burne.

"I went out upon it last night in the darkness, and followed it for a couple of miles, excellency. It is dangerous, but with care we can get safely along."

"You have quite cleared the passage, then?" said the professor.

"Right to the mouth, effendi. There, so as not to excite notice, I have only left a hole big enough to crawl from. Not that anyone could see, except from the mountain on the other side, and nobody is ever there."

"When do we go, then?" said Lawrence eagerly.

"If their excellencies are willing, to-morrow night," said Yussuf. "Every hour I am expecting to see the messenger return, and you, gentlemen, forced to agree to some terms by which in honour you will be bound to pay heavy amounts, and then it will not be worth while to escape."

"I say, look here, Yussuf," said Mr. Burne, "are you real or only sham?"

Yussuf frowned slightly.

"Your excellency never trusted me," he replied proudly.

"I did not at first, certainly," said the old lawyer. "I'll go so far as to say that in the full swing of my suspicions I was almost ready to think that you had been playing into the brigands' hands and had sold us."

"Oh, Mr. Burne!" cried Lawrence reproachfully.

"You hold your tongue, boy. You're out of court. You haven't been a lawyer for nearly forty years; I have."

"I have tried hard to win Mr. Burne's confidence," said Yussuf gravely. "I am sorry I have failed."

"But you have not failed, my good fellow," cried the old lawyer. "I only say, Are you a real Turk or a sham?"

"Will your excellency explain?" said Yussuf with dignity. "I speak your tongue, and understand plain meanings, but when there are two thoughts in a word I cannot follow."

"I mean, my dear fellow, you so thoroughly understand the thoughts and ways of English gentlemen that it is hard to think you are a born Turk."

"Oh!" said Yussuf smiling. "I have been so much with them, excellency, and—I have tried to learn."

"There's a lesson for you, Lawrence," said the professor smiling. "Well, then, Yussuf, to-morrow night."

"Yes, excellency."

"Then, had we not better tell the Chumleys?"

Yussuf was silent for a few moments.

"I am sorry about them," he said at last. "We

cannot leave them behind, for it would mean their death; but if we fail in our escape, it will be through them. No, excellency, say no word till we are ready to start, and then say, Come!'"

"You are right, Yussuf," said Mr. Burne. "That woman would chatter all over the place if she knew: say nothing, and we must make the best of them. But I say, isn't it turning very cold?"

"Yes, excellency, we are high up in the mountains. There is no other place so high as this, and if we do not go soon the winter will be upon us."

"Winter? not yet," said the professor.

"Your excellency forgets it is winter in the mountains when it may be only autumn in the plains."

CHAPTER XLII.

A SAD FAILURE.

T last!

The Chumleys were fast asleep; the wood fire had burned down into a faint glow that played over the white ashes, and the air seemed to be piercingly cold.

The guards had looked in according to their custom, and then proved how cold it was by stopping by the fire for about a quarter of an hour, talking in a low tone together before going out.

The provisions, principally bread and raisins, were taken out of Yussuf's hiding-place, where he kept the worsted rope, and this latter he wore twisted round his chest, beneath his loose garment, ready in case it might be wanted. The food was made into six packages, and each took his load, leaving two for the Chumleys, and now a short conversation ensued about Hamed, whom they had only seen once since their imprisonment. For the driver had been sent to another part of the old ruins with the horses.

The professor was saying that they ought to try and get Hamed away with them; but Yussuf declared it would be impossible, and said that as a compatriot he was perfectly safe.

Under these circumstances it was decided to leave him; and now, all being ready, Lawrence was deputed to awaken the Chumleys, and bid them rise and follow.

"How do you feel, my lad?" said the professor, with his lips to Lawrence's ears.

"Nervous, sir."

"No wonder. It seems cruel to have to leave so much behind, but never mind. Now, Burne, are you strung up?"

"Yes, quite," was the reply.

"Ready, Yussuf?"

"Yes, excellency, and mind, once more, all are to follow me close under the walls. Not a word is to be spoken."

"But you will pause for a few minutes in the subterranean passage," whispered the professor. "I must see that."

"You will have ample time, excellency. Now, Lawrence effendi, awaken your friends."

Lawrence drew a long breath, and stooping down, laid his hand upon Mr. Chumley's shoulder.

"Don't!" was the gruff response.

"Mr. Chumley, wake up. Hush! Don't speak."

"Eh, what? Time to get up. Why don't you pull aside the rug?"

"Hush, sir! Wake up."

"Eh, what? Is my wife ill?"

"No, no. Are you awake now?"

"Awake? Yes, of course; what is it?"

"We have a way open to escape. Wake your wife. Tell her not to speak."

"But she will. Oceans!" said the little man sadly.

"She must not speak. Wake her; tell her there is a way of escape, and then you two must carry these parcels of food, and follow in silence."

"I say, Lawrence, old man, is it real?" he whispered.

"Quite! Quick! You are wasting time."

"But won't they shoot at us?"

"Not if you are both silent," whispered Lawrence; and creeping on all-fours the little man reached over, awakened his wife, and communicated the news.

To the surprise of all she woke up quite collected, grasped the idea at once, and rose to her feet. Then putting on her head-dress, and throwing a shawl over her shoulders and securing the ends,—

"I am ready," she said.

"Bravo!" whispered the professor. "Now, silence, for we have to pass the guards."

"But where are we going?" said Chumley.

"Chumley! Oh, that tongue!" whispered his wife.

"Silence!" said Yussuf decidedly; and then after a pause, "Ready?"

There was no reply, and taking this for consent, he bade the professor come last, after holding the rugs aside till all had passed, and then he stepped out, and stepped back again, for a piercingly cold breath of air had darted into the prison.

"It is snowing," he said in a low whisper.

"Well?" said Mr. Burne, "we are going down from the mountain, and we shall leave it behind, shall we not?"

"Yes, perhaps," said the guide, in a doubting manner. "Shall we risk it?"

"Yes, certainly," said Mr. Preston. "We must go now."

"It is well," said Yussuf, and he stepped out, the others following in his steps; but when it came to Lawrence's turn, to his intense surprise he found that his feet sank deep in the softly gathering flakes. He looked to his left as he kept on by the wall; but the guards were not visible though their voices could be heard, and it was evident that they had sheltered themselves among some stones where they were gossiping together.

Not a sound was heard but the rush of wind as the little party crept on—their footsteps were effectually muffled, and in a few minutes they were beyond the hearing of the guards, even had they spoken; but they had to keep close together, for the drifting snow was blinding, and hid their footprints almost as soon as they were formed.

Away to their left lay the ruins which formed the robbers' town, and farther away, and still more to the left, lay the way to the entrance, where there was quite a grand room, and a goodly fire burned; but the fugitives could only see snow: the air was thick with it, and they kept on until Yussuf stopped so suddenly that they struck one against the other.

"What is it?" said Lawrence, who was next to him now, the Chumleys having asked him to go before them.

"I have lost my way," said Yussuf angrily; "the snow has deceived me. The old temple should be here."

"Well, here it is," said Lawrence, who had stretched out his hand. "Here is one of the columns."

"Ha!" ejaculated Yussuf; "good boy! Yes, the fourth; I know it by this broken place in the side. Two more steps and we are in shelter."

It was a proof of his admirable powers as a guide to have found the way in the midst of the blinding snow, but no one thought of that. Every mind was strained to the greatest pitch of tension; and when Yussuf led the way into the old temple, and the footsteps were heard upon the marble floor, Mr. Burne started and thought that their pursuers were upon them.

"Here is the place," said Yussuf. "Lawrence effendi," he continued as he raised the stone, "you know the way; go first and lead. I must come last and close the stone, so that they may not know the way we have come."

"Is there any danger?" said Mrs. Chumley excitedly.

"None at all," replied Lawrence. "It is only to walk down some rough steps."

She said no more, but let herself be helped down through the opening, and in five minutes they were all in what seemed to be quite a warm atmosphere, waiting in the intense darkness while Yussuf carefully closed the stone.

"There is nothing to mind," said Lawrence. "I have been all the way down here, and I will tell you when the steps end and the rough slopes begin."

He spoke aloud now, in quite a happy buoyant manner which affected the rest, and their spirits rose still higher when Yussuf suddenly struck a match and lit the lamp which his forethought had provided.

This done they stood in the rugged arched passage to shake off the clinging snow with which they were covered, and with spirits rising higher still the whole party followed Yussuf, who, lamp in hand, now went to the front.

"I should like to stop here for an hour or two to examine this roofing and the steps," said the professor. "Pre-Roman evidently. We have plenty of time, have we not?"

"Effendi, it would be madness," cried Yussuf angrily. "Come on!"

"I have done, and you are master of the situation," said the professor quietly; while Mr. Burne burst into a laugh, took snuff, and then blew his nose, so that it echoed strangely along the passage.

"Effendi!" cried Yussuf reproachfully.

"Tut-tut!" exclaimed the old lawyer. "I thought we were safe."

"How much farther have we to go?" said Mrs. Chumley at last.

"We are at the bottom," replied Yussuf. "Mind, there are stones here. You must mind or you will hurt yourselves, and the wind will put out the lamp directly. There is an opening here, and when I have thrust out a stone or two we shall be on a rocky path. You will all follow me closely. Better take hold of hands; then, if one slips, all can help."

But the wind did not blow out the lamp; and as they stood watching Yussuf creep along a narrow horizontal passage the light shone upon the dazzling snow which had filled up the hole, and after thrusting at it for a few minutes and scraping it down their guide desisted and crept back.

"I feared this," he said sadly.

"Feared! Feared what?" cried Mr. Burne.

"The snow, effendi. The way is blocked; the snow must be drifting down from the mountains and falling in sheets."

"But it will not last, man?"

"Perhaps for days, excellency; and even if the hole were open, I see it would be utter madness to brave the dangers of that shelf of rock in the face of this storm."

"Oh, nonsense!" cried Mr. Burne; "let's go on. We cannot get back."

"His excellency does not know the perils of a mountain snow-storm or he would not say this. Suppose that we could force our way out through that snow, how are we to find the buried path with a precipice of a thousand feet below? No, excellencies, we are stopped for the present and must get back."

"How unfortunate!" cried the professor; "but Yussuf is right—we must return and wait for a better time. Can we get back unseen?"

"We must try, excellency; but even if we are caught, it will not be till after we are out of the passage and the stone is down. This must be kept a secret."

The way back did not seem long. The stone was closed, and, low-spirited and disheartened, they crossed the rugged floor of the old temple and stood once more amid the snow, which had already fallen knee-deep and in places drifted far deeper. But, in spite of the confusion caused by what answered to intense darkness, Yussuf led them straight to the prison-hall, and then close under its walls till the rug yielded to his hand, and as he drew it aside quite a pile of snow crumbled into the well-warmed place and began to melt.

They were safely back without discovery; and there

was nothing left but to shake off the clinging snow, and, after hiding their packages, try to rid themselves of their disappointment in sleep.

CHAPTER XLIII.

THE WINTRY GUARDIAN.

OR four days the snow fell incessantly. The aspect of the whole place was changed, and it was only with difficulty that the appointed guards managed to bring provisions to the prisoners.

Fortunately an ample supply of fuel was stacked by the door, so that a good fire was kept; but on the fourth day no food was brought whatever, and but for the store they had in concealment matters would have looked bad, for there was no knowing how much longer the storm would last.

But on the fifth day the sun shone out brilliantly, and the brigands and their wives were all busy with shovels digging ways from place to place; and when at last the prison-hall was reached it was through a cutting ten feet deep, the snow being drifted right up to the top of the lofty door.

The scene was dazzling; the ruins piled up with the white snow, the mountains completely transformed as they glittered in the sun, and above all the sky seemed to be of the purest blue.

The cold was intense, but it was a healthy inspiriting cold, and the disappointment and confinement of the past days were forgotten as the glorious sunshine sent hope and life into every heart.

In the course of the day the chief came, bringing with him piled on the shoulders of a lad more rugs and fur coats for his prisoners; and a long conversation ensued, in which he told them through Yussuf that he expected his messengers would have been back before now, but they had probably been stopped by the snow, and they must wait patiently now for their return.

A further conversation took place at the door between the chief and Yussuf, and then the former departed.

"Well, Yussuf," said Mr. Preston anxiously; "what does he say? Not execution yet from his manner?"

"No, excellency; it is as I feared."

"Feared?" cried Mrs. Chumley excitedly; "are we to be kept closer prisoners?"

"No, madam; you are to have greater freedom now."

"Freedom?" all chorused.

"Yes," said Yussuf; "you are to be at liberty to go where you please in the old city, but it will not be far, on account of the snow."

"And outside the town?" said the professor.

"Outside the town, excellency," said Yussuf sadly. "You do not realize that we had a narrow escape that night."

"Escape?"

"Yes, of being destroyed; the snow everywhere is tremendous. Even if no more comes, we shall be shut in here, perhaps, for months."

"Shut in?"

"Yes; the mountains are impassable, and there is nothing for it but to submit to fate."

"But the snow will soon melt in this sunshine."

"No, excellency, only on the surface, unless there is

a general thaw. You forget where we are, high up in the Dagh. Even where the snow melts, it will freeze every night, and make the roads more impassable. As to our path by the side of the precipice it will not be available for months."

There was a serious calm in Yussuf's words that was most impressive. It seemed so hard, too, just as they had been on the point of escaping, for the winter to have closed in upon them so soon, and with such terrible severity; but that their case was hopeless seemed plain enough, for the guards were withdrawn from their door, and in the afternoon they relieved the tedium of their confinement by walking along the cuttings that had been made.

On every hand it could be seen that the brigands were accustomed to such events as this; firing and food had been laid up in abundance, and whether the winter, or an enemy in the shape of the government troops, made the attack, they were prepared.

"There is nothing for it, Lawrence, but to accept our position, I suppose," said the professor.

"No," said Mr. Burne, who overheard the remark; "but suppose my snuff does not hold out, what then?"

Before anyone could answer, he made a suggestion of his own.

"Necessity is the mother of invention," he said. "I should have to bake some of this Turkish tobacco and grind it between stones.

Then a week glided away, and during that time, being left so much to their own devices, the brigands keeping in the shelter of their homes, the professor visited the ancient passage with Yussuf, and carefully explored it.

(348)   Y

"Ancient Greek," he said when he returned, "like the greater part of this old city. Some of it has been modernized by the Romans, but that passage is certainly ancient Greek, about—"

"But the way out—the way to escape, Mr. Preston," said Mrs. Chumley eagerly, "surely that is of more consequence than your dates."

"To be sure, yes; I forgot, ma'am. Yussuf made a careful investigation of the mouth of the passage where it opens upon the side of the precipice; in fact, I went out with him. The track is many feet deep in snow, and it would be utter folly to attempt to escape."

"Oh, dear me!" sighed Mrs. Chumley.

"We must bear our lot patiently till the first thaw comes, and then try and make our way over the mountains."

These were the words of wisdom, and for long weary weeks the prisoners had to be content with their position. The brigands did a little snow-cutting, and then passed the rest of their time sleeping by the fires they kept up night and day. Food was plentiful, and the chief behaved civilly enough, often paying his prisoners a visit, after which they were entirely left to their own resources.

"We ought to be low-spirited captives," Mr. Burne used to say, as he beat his hands together to keep them warm; "but somehow nobody seems very miserable."

And this was a fact, for every day the professor kept them busy with shovels digging away the snow from some piece of ruin he wished to measure and draw, while after the chief had been, and noted what was done, he said something half contemptuously to his men, and no interference took place.

Day after day, with a few intervals of heavy snow

and storm, the dazzling sunshine continued, with the brilliant blue sky, and the mountains around looking like glistening silver.

Everywhere the same deep pure white snow, in waves, in heaps, in drifts, and deep furrows, silvery in the day, and tinged with rose, purple, scarlet, and gold as the sun went down.

They were so shut in that an army of men could not have dug a way to them; and, knowing this, the brigands dropped into a torpid state, like so many hibernating bears, while the professor's work went on.

"Do you know, Lawrence," he said one day, laying down his pencil to rub his blue fingers, "I think I shall make a great book of this when I have finished it. I have got the castle done, the principal walls, the watch-towers and gates, and if there was not so much snow I should have finished the temple; but, bless my heart, boy, how different you do look!"

"Different, sir!" said Lawrence laughing. "Oh, I suppose the wind has made my nose red."

"I did not mean that: I meant altogether. You look so well."

Lawrence had been handling a shovel, throwing snow away from the base of an old Greek column, and he smiled as he said:

"Oh, I feel very well, sir."

He need not have spoken, for the mountain air had worked wonders. Nature was proving the best doctor, and the enforced stay in that clear pure air, with the incessant exercise, had completely changed the lad.

CHAPTER XLIV.

THE EVASION.

HREE months had passed away, and though the hopes of the prisoners had been raised several times by the commencement of a thaw, this had been succeeded again and again by heavy falls of snow, and by repeated frosts which bound them more closely in the stronghold.

But at last the weather completely changed. The wind came one day cloud-laden, and with a peculiar sensation of warmth. Thick mists hid the mountain tops, and filled up the valleys, and a few hours later the professor and his companions had to make a rush for the shelter of the great hall that was their prison, for a terrific downpour commenced, and for the next fortnight continued almost incessantly.

The change that took place was astounding; the mountain sides seemed to be covered with rills, which rapidly grew, as they met, into mountain torrents, which swirled and foamed and cut their way through the dense masses of snow, till they were undermined and fell with loud reports; every now and then the loosened snow high up began to slide, and gathered force till it rushed down as a mighty avalanche, which crashed and thundered on its course, bearing with it rock and tree, and quite scraping bare places that had been covered with forest growth.

At first the prisoners started up in alarm as they heard some terrible rush, but where they were placed was out of danger; and by degrees they grew used to the racing down of avalanche, and the roar of the leaping and bounding torrents, and sat talking to Yussuf all through that wet and comfortless time about the probabilities of their soon being able to escape.

"The snow is going fast," he said; "but for many days the mountain tracks will be impassable. We must wait till the torrents have subsided: we can do nothing till then."

Nearly four months had passed, since they had met the brigands first, before Yussuf announced that he thought they might venture to make a new attempt. The snow had pretty well gone, and the guards were returning to their stations at the great gate. There was an unwonted hum in the settlement, and when the chief came he seemed to take more interest in his prisoners, as if they were so many fat creatures which he had been keeping for sale, and the time had nearly come for him to realize them, and take the money.

In fact, one day Yussuf came in hastily to announce a piece of news that he had heard.

The messengers were expected now at any moment, for a band of the brigands had been out on a long foraging excursion, and had returned with the news that the passes were once more practicable, for the snow had nearly gone, save in the hollows, and the torrents had sunk pretty nearly to their usual state.

"Then we must be going," said Mr. Burne, "eh?"

"Yes, effendi," said the guide, "before they place guards again at our door. We have plenty of provisions saved up, and we will make the attempt to-night."

This announcement sent a thrill through the little party, and for the rest of the day everyone was pale with excitement, and walked or sat about waiting eagerly for the coming of night.

There was no packing to do, except the tying up of the food in the roughly-made bags they had prepared, and the rolling up of the professor's drawings—for they had increased in number, the brigand chief having, half-contemptuously, given up the paper that had been packed upon the baggage horses.

Mr. Preston was for making this into a square parcel, but Yussuf suggested the rolling up with waste paper at the bottom, and did this so tightly that the professor's treasure, when bound with twine, assumed the form of a stout staff—"ready," Mr. Burne said with a chuckle, "for outward application to the head as well as inward."

All through the rest of that day the motions of the people were watched with the greatest of anxiety, and a dozen times over the appearance of one of the brigands was enough to suggest that suspicion had been aroused, and that they were to be more closely watched.

But the night came at last—a dark still night without a breath of air; and as, about six o'clock as near as they could guess, everything seemed quiet, Yussuf went out and returned directly to say that there were no guards placed, and that under these circumstances it would be better to go at once. No one was likely to come again, so they might as well save a few hours and get a longer start.

This premature announcement startled Mrs. Chumley, so that she turned faint with excitement, and unfortunately the only thing they could offer her as a restorative was some grape treacle.

This stuff Chumley insisted upon her taking, and the annoyance roused her into making an effort, and she rose to her feet.

"I'm ready," she said shortly; and then in a whisper to her husband, "Oh, Charley, I'll talk to you for this."

"Silence!" whispered Yussuf sternly. "Are you all ready?"

"Yes."

"Then follow as before, and without a word."

He drew aside the rug, and the darkness was so intense that they could not see the nearest building as they stepped out; but, to the horror of all, they had hardly set off when a couple of lanterns shone out. A party of half a dozen men, whose long gun-barrels glistened in the light, came round one of the ruined buildings, and one of them, whose voice sent a shudder through all, was talking loudly.

The voice was that of the chief, and as the fugitives crouched down, Yussuf heard him bid his men keep a very stringent look-out, for the prisoners might make an attempt to escape.

Yussuf caught Lawrence's hand and drew him gently on, while, as he had Mrs. Chumley's tightly grasped, she naturally followed, and the others came after.

"Quick!" whispered Yussuf, "or we shall be too late."

The darkness was terrible, but it was in their favour, so long as they could find the way to the old temple; and they needed its protection, for they had not gone many yards among the ruins before there was an outcry from the prison, then a keen and piercing whistle twice repeated, and the sounds of hurrying feet.

Fortunately the old temple lay away from the inhabited portion; and as they hurried on, to the great joy of all they found that the chief and his men were not upon their track, but were hurrying toward the great rock gates, thus proving at once, so it seemed, that they were ignorant of any other way out of the great rock fortress.

Once or twice Yussuf was puzzled in the darkness, but he caught up the trail again, and in a few minutes led them to the columned entrance of the temple, into whose shelter they passed with the noise and turmoil increasing, and lights flashing in all directions.

"Hadn't we better give up?" said Mr. Chumley, with his teeth chattering from cold or dread.

"Give up! What for?" cried Mr. Burne.

"They may shoot us," whispered the little man. "I don't mind, but—my wife."

"Silence!" whispered Yussuf, for the noise seemed to increase, and it was evident that the people were spreading all over the place in the search.

As Yussuf spoke he hurried them on, and in a minute or two reached the stone that led to the passage in the rift.

It was quite time he did, for some of the people, who knew how they had affected that place, were making for the temple.

But Yussuf lost no time. He turned up the stone in an instant, and stood holding it ready.

"Go first, Lawrence effendi," he whispered; "help Lady Chumley and lead the way."

Lawrence dropped down at once, and Mrs. Chumley followed with unexpected agility; then Chumley, Mr. Burne, the professor; and as Yussuf was following, lights flashed through the old building, and lit up the roof.

Fortunately the ruins of the ancient altar sheltered the guide, as he stepped down and carefully lowered

THE ESCAPE OF THE PRISONERS FROM THE BANDITS.

the stone over his head as he descended; and so near was he to being seen that, as the stone sank exactly into its place, a man ran over it, followed by half a dozen more, their footsteps sounding hollow over the fugitives' heads.

Meanwhile Lawrence hurried Mrs. Chumley down, the others following closely, till the bottom of the steps and slopes was reached, and the cool night air came softly in through the opening.

There they stopped for Yussuf to act as guide; but, though his name was repeated in the darkness again and again, there was no answer, and it soon became evident that he was not with the party.

"We cannot go without him," said Mr. Preston sternly. "Stop here, all of you, and I will go back and try to find him." But there was no need, for just then they heard him descending.

"I stopped to listen," he said. "They have not yet found our track, and perhaps they may not; but they are searching the temple all over, for they have found something, and I don't know what."

"My bag of bread and curd!" said Mr. Chumley suddenly. "I dropped it near the door."

"Hah!" ejaculated Yussuf; but no one else said a word, though they thought a great deal, while Mr. Chumley uttered a low cry in the darkness, such a cry as a man might give who was suffering from a sharp pinch given by his wife.

The next moment the guide passed them, and they heard him thrust out a stone, which went rushing down the precipice, and fell after some moments, as if at a great distance, with a low pat. Then Yussuf bade them follow, and one by one they passed out on to a narrow rocky shelf, to stand listening to the buzz

of voices and shouting far above their heads, where a
faint flickering light seemed to be playing, while they
were in total darkness.

"Be firm and there is no danger," said Yussuf; "only
follow me closely, and think that I am leading you
along a safe road."

The darkness was, on the whole, favourable, for it
stayed the fugitives from seeing the perilous nature
of the narrow shelf, where a false step would have
plunged them into the ravine below; but they followed steadily enough, with the way gradually descending. Sometimes they had to climb cautiously
over the rocks which encumbered the path, while twice
over a large stone blocked their way, one which took
all Yussuf's strength to thrust it from the narrow path,
when it thundered into the gorge with a noise that
was awful in the extreme.

Then on and on they went in the darkness, and
almost in silence, hour after hour, and necessarily at a
very slow pace. But there was this encouragement,
that the lights and sounds of the rock-fortress gradually died out upon vision and ear, and after turning a
sharp corner of the rocks they were heard no more.

"I begin to be hopeful that they have not found
out our way of escape," said Mr. Preston at last in a
cheerful tone; but no one spoke, and the depressing
walk was continued, hour after hour, with Yussuf
untiringly leading the way, and ever watchful of perils.

From time to time he uttered a few words of warning, and planted himself at some awkward spot to give
a hand to all in turn before resuming his place in front.

More than once there was a disposition to cry halt
and rest, for the walk in the darkness was most ex-
hausting; but the danger of being captured urged all

to their utmost endeavours, and it was not till daybreak, which was late at that season of the year, that
Yussuf called a halt in a pine-wood in a dip in the
mountains, where the pine needles lay thick and dry;
and now, for the first time, as the little party gazed
back along the faint track by which they had come
through the night, they thoroughly realized the terrible nature of their road.

"Everyone lie down and eat," said Yussuf in a low
voice of command. "Before long we must start again."

He set the example, one which was eagerly followed,
and soon after, in spite of the peril of their position
and the likelihood of being followed and captured by
the enraged chief, everyone fell fast asleep, and felt as
if his or her eyes had scarcely been closed when, with
the sun shining brightly, Yussuf roused them to continue their journey.

The path now seemed so awful in places, as it
ran along by the perpendicular walls of rock, that
Chumley and Lawrence both hesitated, till the latter
saw Yussuf's calm smile, full of encouragement, when the
lad stepped out firmly, and seeing that his wife followed,
the little man drew a long breath and walked on.

Now they came to mountain torrents that had to be
crossed; now they had to go to the bottom of some
deep gorge; now to ascend; but their course was
always downwards in the aggregate, and at nightfall,
when Yussuf selected another pine-wood for their
resting-place, the air was perceptibly warmer.

The next morning they continued along the faintly-
marked track, which was kept plain by the passage of
wild animals; but it disappeared after descending to a
stream in a defile; and this seemed to be its limit, for
no trace of it was seen again.

For six days longer the little party wandered in the mazes of these mountains, their guide owning that he was completely at fault, but urging, as he always led them down into valleys leading to the south and west, that they must be getting farther away from danger.

It was this thought which buoyed them up during that nightmare-like walk, during which they seemed to be staggering on in their sleep and getting no farther.

It seemed wonderful that they should journey so far, through a country that grew more and more fertile as they descended from the mountains, without coming upon a village or town; but, though they passed the remains of three ancient places, which the professor was too weary to examine, it was not until the seventh day that they reached a goodly-sized village, whose headman proved to be hospitable, and, on finding the state to which the travellers had been reduced and the perils through which they had passed, he made no difficulty about sending a mounted messenger to Ansina, ninety miles away, with letters asking for help.

CHAPTER XLV.

HOMEWARD BOUND.

EXHAUSTED as the travellers were, sleep, good food, and the soft sweet air soon restored them, and they were ready to continue their journey long before their messenger returned, to bring faithfully the means for a fresh start, with fresh ponies, and the necessaries they required, though these were hard to obtain in so out-of-the-way a place.

The weather was threatening as they started at last for Ansina, the Chumleys electing to accompany them. In fact, on parting, their host, who had been amply recompensed for his kindness, warned them to hasten on to the port, for snow, he said, would fall before the week was out, and then the famished wolves would descend from the mountains and the plain become dangerous.

The advice was readily taken, for all were quite satisfied that their travels in Asia Minor would be better ended for the present.

In this spirit they made the best of their way to the port, where they arrived with the snow falling slightly, though high up in the mountains there was a heavy storm. They took up their quarters at the best hotel in the place, and could have gone on at once by the steamer from Beyrout, but at Lawrence's wish the departure was put off till the coming of the next boat, a fortnight later.

"You do not feel so well?" said Mr. Preston anxiously.

"Eh, what, not so well?" cried Mr. Burne, turning to look at Lawrence. "Look here, don't say that. I thought we had cured him."

"Oh, I'm quite well and strong," cried Lawrence quickly.

"But you seem so dull," said the professor.

Lawrence did not answer, but turned away his head.

"I wish we had gone on," said Mr. Preston anxiously. "There would have been good medical advice on board."

"No, no, I am not ill," said Lawrence; and then in a broken voice, he cried excitedly, "I wanted to put it off as long as I could."

"What! going home, my dear lad?" said Mr. Burne eagerly "You are afraid of our climate again. Then let's stay."

"No, no; it was not that," said Lawrence. "I—I
—there, I must say it. Yussuf has—has been such a
good fellow, and we shall have to say good-bye at
Smyrna."

The professor was silent for a few minutes.

"Perhaps not for always," he said at last. "Yes: he
has been a thoroughly good fellow, and I, for one,
should like to come out and have another trip with
him. What do you say?"

"Yes, yes," cried Lawrence eagerly; and he rushed
out of the room, to be seen the next minute holding
on by the grave-looking Turk's arm and telling him
the news.

"Look at that," whispered Mr. Burne to the professor, as he eagerly watched Yussuf's countenance.
"Now, if ever anyone tells me in the future that the
Turks always hate the Christians, I can give him an
instance to the contrary."

The time soon glided by for the coming of the next
boat, and in due course they landed at Smyrna, where
the parting with Yussuf was more that of friends and
friend, than of the employer and employed.

"If you do come out again, excellencies, and I am
living, nothing shall stay me from being your faithful
guide," he said, as he stood at the gangway of the
steamer; and as for you, Lawrence effendi, may the
blessings spoken of by the patriarchs be with you in
your goings out and comings in, and may the God of
your fathers give you that greatest of his blessings,
health."

Lawrence did not speak, but clung to the faithful
hand till the Turk descended into the boat; and he
then stood gazing over the gangway till the grave,
thickly-bearded countenance grew less and less and at
last died from his sight.

The little party landed at Trieste, where they parted

from the Chumleys, who were going home; but Lawrence and his friends, after repairing the damages to
their wardrobes, went by rail to Rome, and made that
their home till the rigour of the English spring had
passed away.

It was one fine morning at the beginning of June,
that a cab laden with luggage stopped at the old home
in Guilford Street, where the door was opened by
Mrs. Dunn, who stared with astonishment at the sturdy
youth who bounded up the steps into the hall, and
then clasped her in his arms.

"Why, my dear, dear boy!" she cried, "I had
brought blankets down to wrap you in, and a warm
bath ready, and asked cook's husband to be in waiting
to carry you upstairs."

"Why, nurse, I could carry you up," cried Lawrence merrily. "How well you look! Ah, Doctor
Shorter."

"Why, you wicked young impostor," cried the
doctor; "here have I neglected two patients this afternoon on purpose to come and attend on you. I came
as soon as nurse Dunn told me she had received the
telegram from Folkestone. Bless my heart, how you
have changed!"

"Changed, sir?" cried Mr. Burne, "I should think he
has changed. He has been giving up physic, and
trusting to the law, sir. See what we have done!"

"Yes, doctor," said the professor, shaking hands
warmly. "I think you may give him up as cured."

"Cured? That he is!" cried the doctor. "Well, live
and learn. I shall know what to do with my next
patient, now."

"And if here isn't Mrs. Dunn crying with vexation,
because she has no occasion to make gruel and mix
mustard plaisters for the poor boy," cried Mr. Burne
banteringly.

"No, no, no, sir," said the old woman sobbing; "it is

out of the thankfulness of my poor old heart at seeing my dear boy once more well and strong."

The doctor took out his note-book, and made a memorandum as Lawrence flung his arms round the tender-hearted old woman's neck; the professor walked to the window; and Mr. Burne whisked out the yellow handkerchief he had worn round his fez, and over which he had made his only joke, that he was so yellow and red, he looked like a fezzan, and blew his nose till the room echoed. After which he was obliged to calm himself with a pinch of snuff.

"Well, Lawrence," said the professor, after they had all dined together. "You remember what you said at Ansina?"

"Yes."

"What do you say now? Would you go through all those wearinesses and risks again if I asked you?"

"Yes, sir, at any time, if Yussuf is to be our guide."

"And so say I," cried Mr. Burne, "if you would have such a cantankerous old man."

"Ah, well," said the professor. "I am not half satisfied. We shall see."

And so it was left.

THE END.